Regenerative Medicine Ethics

Linda F. Hogle
Editor

Regenerative Medicine Ethics

Governing Research and Knowledge Practices

Springer

Editor
Linda F. Hogle
University of Wisconsin—Madison
Madison, WI, USA

ISBN 978-1-4614-9061-6 ISBN 978-1-4614-9062-3 (eBook)
DOI 10.1007/978-1-4614-9062-3
Springer New York Heidelberg Dordrecht London

Library of Congress Control Number: 2013953877

Printed on acid-free paper

Springer is part of Springer Science+Business Media (www.springer.com)

Preface

Motivation for this book came from two very different things: frustration and a desire for reciprocity. Frustration, because it is difficult to find teaching materials on research ethics that are both specific enough to be useful to regenerative medicine researchers and yet robust enough to be applicable to a broader audience of those who deal with similar issues in scientific practice, particularly those related to the use of human and animal cells in research. Frustration also comes from the narrow way in which ethical and social issues in regenerative medicine have been defined. The subject of much of ethical and policy writing about regenerative medicine, while important, is often both disproportionate and out of sync with what researchers, clinicians, patients, and others face in the lab, the classroom, the boardroom, in exchanges within scientific communities, or in everyday decisions about health care. Readers are accustomed to the debates that have dominated stem cell research to date, filtered through historical and political lenses, but there are emerging issues in the way regenerative medicine is being practiced and organized that go begging for analysis and discussion. On the horizon are surprising changes in our ability to manipulate biology and disruptions to the way we have organized and funded the production of knowledge, all of which need careful consideration. But there are also gray areas and "unknown unknowns" that come up in routine practice and need guideposts for good judgment. This book thus has dual aims: to bring attention to a few salient emerging issues, as well as to provide fundamental information that will be useful for both experienced readers and those new to the field. In addition, I hope that readers will see that social, ethical, policy, and technological issues are mutually constitutive. That is, governance structures, conventions for sequestering or sharing knowledge, research priorities and financing, new scientific theories, and technical tools are all informed by history, politics, and culture and all are involved in the dynamic processes involved in producing knowledge.

Then there's reciprocity. Much of my research and teaching has been grounded in work previously supported by the National Science Foundation and the Greenwall Foundation. I am deeply appreciative for the opportunities this funding provided. It has also been my privilege to work with, learn from, and teach many researchers working in regenerative medicine. I am grateful to the many stem cell and tissue engineers, biologists, materials scientists, regulatory officers, and others who have not only patiently endured my many questions over the years but also engaged with me in thinking through the many challenges and paradoxes inherent in regenerative medicine research. This book is for you, and for all of the researchers, policy-makers, program managers, and students who must navigate constantly changing environments. I would also like to thank research assistants Jessica Von Reyn and Catherine Turng, who made this challenging book manageable, as well as the many students who have made the journey so rewarding.

Madison, WI, USA Linda Hogle

Contents

Part I
Foresight, Hindsight and Peripheral Vision: Navigating Contemporary Legal, Ethical and Policy Issues in Regenerative Medicine

Chapter 1
Contemporary Issues in Regenerative Medicine Research Ethics and Governance: An Overview

Linda F. Hogle

1.1 Introduction

It has been 15 years since the first successful derivation of human embryonic stem cell (hESC) lines. Together with insights from tissue engineering over the past 3 decades, the capability to reprogram differentiated somatic cells, and many other innovations in techniques and materials, regenerative medicine has become a vibrant, interdisciplinary field. Regenerative medicine (RM) now involves many kinds of cells and bioreactive materials and a wide variety of techniques, including contributions from synthetic biology, gene editing, nanotechnology, computational biomechanics, data analytics, optogenetics, and much more. At the same time, translational researchers are coping with complex issues of scaling up human cell and tissue culture, biomanufacturing, and therapeutic delivery systems. As the field matures, new questions arise about how best to manage the knowledge, data, and materials being produced and whether existing policy, law, and ethical oversight systems suit such rapidly evolving science. At this pivotal point, innovations in infrastructure and institutional systems—including governance—are needed as much as those at the lab bench. Yet much of the literature on ethics, policy, and legal issues has not matured to meet the needs of the field as it is today. The lion's share of the many volumes published about stem cell ethics, policy, and legal issues in bioethics and scientific literature has focused on a narrow range of topics: controversies around the use of embryos, cloning, and the problems of policy

L.F. Hogle (✉)
University of Wisconsin—Madison, 1300 University Avenue, 1135
Medical Sciences Building, Madison, WI 53706, USA
e-mail: lfhogle@wisc.edu

L.F. Hogle (ed.), *Regenerative Medicine Ethics: Governing Research and Knowledge Practices*, DOI 10.1007/978-1-4614-9062-3_1,

interoperability across borders.[1] The field has moved on; ethical analysis of research and knowledge practices should too.

Whether unlocking the molecular mechanisms of disease, studying cell microenvironments, building microfabrication systems, or working with clinical and industrial partners to commercialize products, there are challenges arising from both the novelty of the science and the institutional environments in which research exists. These include deciding how to deal with sensitive or proprietary data, creating interoperability of guidelines and protocols across jurisdictions, knowing best to manage collaborations across disciplines and sectors (academic, governmental, and industry), and negotiating legal and policy changes. These emerging knowledge practices have often received less attention but will significantly affect planning and execution of responsible research. There are also ongoing research ethics issues as well: both experienced researchers and those new to the field will encounter a number of issues in everyday practice in the lab, when they submit articles to a journal or proposals to a funding agency, or when they conceive of novel approaches to replace or restore human tissue function. *Regenerative Medicine Ethics: Governing Research and Knowledge Practices* addresses some of these governance issues facing contemporary RM research now and in the near future.

The book was created primarily for regenerative medicine researchers, project managers and strategic planners, and those working in policy or law, who are preparing for next-generation technologies and experimental systems. While the discussions and illustrations in the book come primarily from stem cell research, tissue engineering, and related areas of research, the issues raised by the authors apply far beyond regenerative medicine and will be of considerable relevance to researchers in genetics and "omics" fields, computational biology and data analytics in science, materials scientists, and more.

Some of these issues have not yet been sufficiently addressed in the ethics literature, are not often formally discussed in scientific conferences, or are not presented in a way that is salient for regenerative medicine research. The book thus aims to provide information with which to prepare for current and upcoming issues in the field. The book further aims to help readers understand why current issues have developed the way they have and what implications continue to exist for their work going forward. Understanding the tensions between competing social goods and seeing the contexts behind policies in different societies will hopefully help researchers to put current and emerging debates into perspective. Critically questioning why some issues rise to the level of being and ethical or legal "problem," while others are ignored or downplayed will help with understanding institutional, social, and historical framings of dilemmas in science policy and ethics.

No single book could be comprehensive, and many emerging issues specific to new techniques and discoveries will not be covered in this volume. For example, there is not a chapter devoted entirely to emerging product safety and efficacy

[1] In a survey of tissue engineering journal articles, for example, embryo use and cloning overwhelmingly dominated the range of potential ethical topics of interest (de Vries et al. 2008).

oversight, although several chapters address specific relevant issues.[2] Most of the contributors find that regulatory issues are still in flux. It appears that the scope of regulatory purview may be changing, as judged by the FDA's recent inclusion of cord blood and autologous transplants into their domain. For reprogrammed cells, there may likely be additional testing required to ensure that haplotype changes do not occur, and there may potentially be new requirements for products using gene editing techniques (Condic and Rao 2008). It is still early to predict what trends will occur until more products are ready to enter trials.

Also, because the aim is to address matters pertaining to current basic and translational research practice and future emerging issues, the book does not detail debates that characterized the early era of embryonic stem cell research, including those about the moral status of the human embryo, cloning, political ideologies prevalent in the early twenty-first century, or other political and ethical concerns. These debates, along with insights from moral philosophy, have been extensively reviewed elsewhere, and additional resources can be found on websites found in Appendix C.[3] Previous issues cannot be ignored, however, since they continue to shape policies and laws. Section 1.1 of this introductory chapter shows how such issues, arising from broader social, cultural, and political concerns, remain unsettled and raise new questions for how we should govern biotechnologies using human biological materials.

Part I of the book addresses emerging issues that researchers are beginning to confront, whether conducting discovery research far from use in humans, preclinical research and human trials, or entrepreneurial science aiming to enter commercial markets. Part II is the "bread and butter" of what researchers need to know for routine laboratory practice, whether conducting discovery science or clinical research at the point of moving products into use in human patients. This introductory chapter provides an overview of chapters, but first begins by reviewing salient legal and policy activities that will have an ongoing effect on regenerative medicine research.

1.2 Legal and Policy Updates

Many of the legal and policy issues facing regenerative medicine continue to have to do with the nature of the materials being used. Two long-standing questions involving definitions of life remain unresolved, even after court cases and legislative

[2] Basic information necessary for the review of RM products in the USA and Europe can be found in Fink (2009) and Faulkner (2009), among others.

[3] A collection of reflections by philosophers and religious leaders appears in Holland et al. (2001). Discussions of issues important in early stages of determining public policy can be found in an archive of the National Bioethics Advisory Commission reports on stem cell research, at http://bioethics.georgetown.edu/pcbe/reports/past_commissions/nbac_stemcell1.pdf. See also Cohen (2007) for US debates and for Europe, Gottweis et al. (2009). Debates about the use of embryos in research in general can also be found in Brock (2010), Singer and Kuhse (1986), and the original NIH Human Embryo Research Panel Report (1994).

debates. The first is whether it should be morally permissible to use human embryos in research, to derive cell lines from embryos, and to destroy embryos in the process of conducting research.[4] The second is whether—or to what extent—entities created through biotechnology and used in regenerative medicine (processed or genetically manipulated cell lines, genes or SNPs, chimeras, cybrids) are products of nature, and how the determination affects classification as patentable entities and products subject to regulation. This section briefly discusses court cases and legislation that may affect RM researchers in the near and long term.

Trends in policymaking environments around the globe are also impacting RM, including initiatives to support translational research and commercialization. The implications of this shift are briefly discussed at the end of this section.

1.2.1 Defining What can be Used in Research: Sherley v Sibelius and Personhood Laws

Some countries have passed laws addressing stem cell research (either directly or through patenting or regulatory legislation), while others such as the USA have no specific law allowing or banning the research but have instead adopted funding or procedural policies that shape what is possible to pursue within national boundaries.[5] Still, legal challenges have attempted to block hESC research specifically or broader uses of embryos in research, with consequences that reverberate across national borders. In the USA, Sherley v Sibelius raised questions about whether hESC research violated a legal provision banning the destruction of embryos for research and using taxpayer funds to do so. At the state level, personhood laws are being proposed which, if passed, will affect a variety of research within that state.

The use of embryos and fetal tissue in research is linked to long-standing cultural beliefs about the beginning of life and how it may be ended, what constitutes respect of persons (and who or what counts as a "person"), and views about the sanctity of life in any form, regardless of sentience, potentiality, or stage of development. However, as Jasanoff argues, such cultural beliefs are part and parcel of the way societies deliberate and govern the life sciences (2004). That is, constitutional and administrative law differs across societies in terms of how matters of "life" and "nature" are adjudicated, and political and social histories around particular kinds

---

[4] Debates about the use of embryos in research in general can also be found in Singer and Kuhse (1986) and the original NIH Human Embryo Research Panel Report (1994). The debates primarily pitted potential therapeutic benefit of the research against cultural and religious meanings of nascent life. There are also important matters related to the derivation of lines from embryos, the timing and manner of request for donation, and disposition of embryos, which are discussed in Chap. 6, Sects. 6.2.2 and 6.3.1 of this volume.

[5] See Appendix A for a listing of laws by country. Chapters 4 and 6 also deal with international laws.

of science and technologies within societies shape the form that law and policies ultimately take. The relative roles of public debate, lobbying efforts from interested participants (including adversaries or proponents, funders, local governments or politicians, in addition to scientists interested in doing research), market dynamics, and other economic drivers all are a part of the negotiation of science governance.

Soon after the 2009 National Institutes of Health (NIH) Guideline revisions lifted restrictions on federal funding for research using stem cells derived from embryos, a formal complaint was brought by Drs. James L. Sherley and Theresa Deisher (adult stem cell researchers) and several antiabortion groups.[6] They argued that the new guidelines violated a ban on the use of federal funds for embryo research. Specifically, a rider attached to Department of Health and Human Services (DHHS) appropriations bills every year since 1995, called the Dickey-Wicker Amendment, prohibits federal funding of research using human embryos in which embryos are destroyed (Section 509, Omnibus Appropriations Act, 2009, Pub. L. 111–8, 3/11/09).[7] The term "embryo" is poorly defined in the statute, and technologies such as the derivation and immortalization of lines from embryonic cells were not anticipated at the time. Neither did it anticipate the predicament of not conducting research on the embryo itself, but rather on cells removed from embryos. The plaintiffs contended that because embryos are destroyed in the course of deriving hESCs, the entire field of hESC research constitutes "research in which human embryos are destroyed." The DHHS, on the other hand, argued that embryos are not destroyed or discarded as part of the funded research projects, which used the cell lines derived from embryos. At stake was whether a distinction could be made between research on embryos in which they are destroyed and research on cells after line derivation for purposes of satisfying the statute. By splitting out the research distinctions, the more difficult question about ambiguous definitions of the beginning of life and what was morally permissible was bypassed.

On August 23, 2010 a District Court granted a preliminary injunction which halted all existing federal funding for hESC research. Researchers had to suspend such research and either shut down experiments, shift to non-embryonic sources, or

---

[6] Cosignatories of the complaint included Nightlight Christian Adoptions (a nonprofit, licensed adoption agency dedicated to protecting and finding adoptive parents for human embryos conceived through in vitro fertilization), the Christian Medical Association, and "all individual human embryos whose lives are now at risk".

[7] The DHHS authorizes funds for the NIH, the major federal funder of research in the USA. Part 45 CFR 46.204 and 207 of Public Health Service Act defines an embryo as: "any organism not protected as a human subject under 45 CFR 46 as of the date of enactment of the governing appropriations act, that is derived by fertilization, parthenogenesis, cloning, or any other means from one or more human gametes or human diploid cells." SEC. 509 of the code specifies that (a) none of the funds made available in this act may be used for (1) the creation of a human embryo or embryos for research purposes or (2) research in which a human embryo or embryos are destroyed, discarded, or knowingly subjected to risk of injury or death greater than that allowed for research on fetuses in utero under 45 CFR 46.208(a)(2) and Section 498(b) of the Public Health Service Act [1](42 U.S.C. 289g(b)) (Title 42, Section 289g(b), United States Code).

find other funds to continue, creating considerable concern about the future of the field. There were also significant impacts beyond the USA with the growing number of international collaborations (Wadman 2010a). In April 2011, an appeals court threw out the preliminary injunction, stating that the original law was ambiguous, and upheld the NIH guidelines allowing federal funding. However, nothing was settled regarding ambiguous definitions of "embryo" or "organism," and the Dickey-Wicker Amendment remained.

The Stem Cell Research Advancement Act (H.R. 4808) introduced in 2010 would have specifically authorized federal funding of hESC research and supersede the rider (Dickey/Wicker Amendment). It was not acted upon in two sessions but was reintroduced in June 2013.[8] Until some action is taken by Congress, the USA remains in a state of legislative ambiguity, leaving the guidelines potentially vulnerable to another challenge. Especially since the three judges in the case all gave differing reasons for their conclusions, there may be an opening for ongoing challenges.

Historians of biology have noted the various ways that embryos and fetuses have been defined and classified over the years as scientific and cultural understandings of the unborn have changed (Dubow 2011; Maienschein 2002; Morgan 2009). There is still considerable confusion stemming from difficulty in staging morphogenesis and the use of imprecise terminology to describe the process (Charo 2001; Downs 2008).[9] This is evident in recent efforts to pass US state laws defining personhood for legal purposes. The proposed laws use differing terminology and stages to describe when a biological entity becomes a "person" (see Appendix B). Some specify "fertilization" or "conception" (but fail to define these terms), while others simply say that an individual person exists from "the beginning of development" (failing to describe what constitutes development) or "at all stages of development." Descriptions of the various bills often use terms "fetus," "embryo," and "unborn child" interchangeably, and increasingly, antiabortion proponents favor the word "preborn" to indicate a continuum of life rather than a change of status before and after the birth event. Wisconsin, for example, has introduced a bill to amend the state constitution to strike the word "born" from the statement about rights of individuals:

Article 1. Declaration of Rights. Equality; inherent rights. SECTION 1. All people are born equally free and independent, and have certain inherent rights; among these are life, liberty and the pursuit of happiness…

---

[8] HR2433 found at http://www.govtrack.us/congress/bills/113/hr2433.

[9] Developmental biologist Karen Downs, among others, argues that fertilization is a process rather than a single instant and that in mammals, individuation does not occur until the point at which embryonic cells become entirely separate from extraembryonic tissue that surround it (2008). This may not sway some philosophers, abortion foes, and some religious traditions which hold that every cell is sacred because of the potential to become a human life. Charo takes up the arguments about potentiality and describes how various policy bodies in the USA have tackled the problem of potential in the context of cloning, which might also be applied to the later development of induced pluripotent stem cells (2001; see also the rebuttal from DiSilvestro 2006).

At the federal level, bills introduced in 2011 (HR 212), again in 2013 (HR 23), define a person as a "one-celled human embryo," which should be granted "all the legal and constitutional attributes and privileges of personhood."[10]

Mississippi was among the first states to attempt passage of a law stating that a fertilized egg qualifies as a human. Many physicians opposed the bill, since it would criminalize terminating a pregnancy for any reason, including ectopic pregnancy. As has been the case in most states where such bills have been proposed, it was defeated, but another effort may be made in 2014, with one suggestion for modification being: "The right to life begins at conception. All human beings, at every stage of development, are unique, created in God's image and shall have equal rights as persons under the law" (http://www.sos.ms.gov/page.aspx?s=7&s1=1&s2=50). Such language lays bare the insertion of ideology from powerful political interests into law.

One of the more eyebrow-raising examples was HR 2036 in Arizona. Most states allow abortion up to 24 weeks, but remarkably, this bill started the clock at 2 weeks after the first day of the last menstrual period (effectively allowing only an 18-week limit). Anyone performing an abortion past this date could be charged with a crime, have his or her license revoked or suspended, and could be held liable for civil penalties if the father of the fetus decides to pursue legal action. Beside the fact that fertilization might not yet have occurred, and even if fertilized, an egg may not implant or survive after implantation, there would be no way to confirm the date of a woman's period other than to ask the woman to provide this information. The bill was thus both technically and pragmatically problematic, not to mention an extreme interpretation of personhood.[11] The bill was signed into law in 2012, but on appeal, it was ruled as unconstitutional in 2013.

In North Dakota, among the states with the most restrictive abortion laws, one version of a personhood bill would have prohibited doctors from disposing of unused embryos after an in vitro cycle. Such a measure would have required unused embryos to be kept frozen indefinitely, which would be costly and emotionally difficult for the couple creating the embryo, and would require burdensome changes to storage facilities. Wisconsin Assembly bill 224 (2013) attempts to prevent the transfer or sale of fetal tissue for any purpose, using this definition: 'Fetal body part' means *a cell, tissue, organ, or other part of, or any material derived from any cell or tissue of, an unborn child*, as defined in s. 12939.75 (1), who is aborted by an induced abortion, as defined in s. 69.01 (13m)" (emphasis added). By addressing

---

[10] Currently, the proposed bill states that every human being shall have all the legal and constitutional attributes and privileges of personhood and defines a human as "each and every member of the species homo sapiens at all stages of life, beginning with the earliest stage of development, created by the process of fertilization, cloning, or its functional equivalent." The term "fertilization" is defined as "the process of a human spermatozoa penetrating the cell membrane of a human oocyte to create a human zygote, a one-celled human embryo, which is a new unique human being." The complete text of the most recent bill at the time of publication is found at http://thomas.loc.gov/cgi-bin/query/z?c113:H.R.23.

[11] It is generally thought that about 70 % of fertilized eggs never implant (Downs 2008). Arizona's legislation was modeled on the "Pain-Capable Unborn Child Protection Act" forwarded by the National Right to Life Committee. Other states are considering similar language.

disposition of embryonic tissues, such bills draw attention to human dignity arguments. Yet adding legislation does not add clarification to deal with the ambiguity of nascent life and may only serve to complicate routine practices in medicine.

While wording differs, and it is difficult to know to what extent legislators understand the terminology they use, such legislation attempts to define a person for legal purposes at the earliest possible stages. Along with related laws being introduced to restrict access to abortions, such moves serve to bolster antiabortion interests. Such social constructions of legal definitions of "person" and "personhood" using biology (or sometimes misstating of biological phenomena) attempt to resolve philosophical and cultural dilemmas but fail. Nevertheless they will have profound implications beyond abortion, including liability and property law, and of relevance to this book, medical research.[12] Clearly all of the current versions will affect hESC research but will additionally affect many other kinds of research using fetal tissue or research on cell lines or other products made from fetal tissue in the past. Many additional kinds of research would not be allowed should such bills become law. For example, HEK (human embryonic kidney cell lines) and WI 38 (made from fetal lung tissue) have been used for decades to make vaccines which have been used in millions of people, to produce human proteins for research, and to study disease mechanisms. The laws would disallow existing research on pregnancy loss and diseases of pregnancy, developmental disorders, and many other kinds of research. When interpreted broadly, they may do more: in vitro fertilization and even some forms of contraception may violate the law (Collins and Crockin 2012).

1.2.2 Legal Definitions of "Nature" and "Manipulation": Prometheus Laboratories, Myriad Genetics, and Regenerative Sciences

If definitions of embryos have been difficult to pin down, so have interpretations of what counts as "natural" in biotechnology, particularly when it comes to making claims to intellectual property over innovations involving genes, cells, or entities arising from the use of embryos. The question of whether—and what sort of—life forms may be patented has been debated for many years, as Noonan notes in Chap. 4. An important recent precedent was *Mayo Collaborative Services v. Prometheus Laboratories, Inc.*, in which a patent claim involved a method for determining

[12] Dubow (2011) provides an excellent analysis of the history of politics around the fetus. She describes long-standing attempts to view fetuses as separate individuals rather than as a neutral entity or a part of the gestational mother. The first US court case regarding a legal claim on behalf of a fetus was Dietrich v Northampton, in 1884, in which a woman sued a city for loss of life when she miscarried after a fall on city property. Definitions of fetuses as "patients" and "deceased minors" are increasingly beginning to appear in state legislatures for purposes beyond abortion. Wisconsin Assembly Bill 160 (2013) states that for purposes of a medical malpractice claim, the term "patient" includes an unborn child and defines "unborn child" as a human being from the time of fertilization to the time of live birth.

metabolite levels of a drug in a patient's blood. The court invalidated the claims, stating that the correlation between metabolite levels and the toxicity and efficacy of the drug followed a natural law. If strictly held, any number of processes in biotechnology (which could be said to use "natural processes") and a number of patents already granted could also be invalidated. Distinctions between methods and materials patents will be important to watch in light of this case as well as a case against Myriad Genetics.

As this book was about to go to press, a landmark Supreme Court overturned decades of interpretations of patent law, ruling that genes cannot be patented because they are products of nature.[13] The company, Myriad Genetics, had patented isolated DNA sequences for the breast cancer genes BRCA1 and BRCA2, methods to determine risk of breast cancer and interpret the findings, and methods to test drugs using the sequence. Kevin Noonan (ibid) details the case and the ultimate ruling, which threw out most of the Myriad patents (see also Liptak 2013; Marshall 2013). Patient groups lauded the decision, since tests for these genes would ostensibly be more affordable for more patients. Others celebrated the principle that human genetic material should not be patented and "owned" by for-profit companies. Proponents of patent protection, however, voiced concern that the ruling may inhibit investment in biotechnology innovation in which human biological materials are used. The outcomes of the case will be important to follow, as the same logic used in Myriad case may well invalidate patents on hES cells and other genetic materials used in regenerative medicine.

An additional case involving an autologous stem cell treatment will likely have profound effects for clinical cell therapies. Regenexx™ is a treatment for joint, bone, tendon, and muscle disorders in which bone marrow and blood samples from a patient are removed and then the stem cells are separated, cultured, and reinjected directly into the injured area in the patient, with the assumption that they will integrate and regenerate tissue. The Food and Drug Administration halted use of the procedure, claiming that the company, Regenerative Sciences, was required to submit a New Drug Application and follow federal Good Manufacturing Practices ("cGMP"). At stake was the question of whether the cells were altered enough through the processing procedure to change their biology. If so, they would be more than "minimally manipulated" and could be classified as a drug. Minimal manipulation is defined by the FDA as "processing that does not alter the relevant biological characteristics of cells or tissues." The firm maintained that this was a medical practice procedure, not a drug, since the cells were from an autologous source and administered within the confines of a clinic. Similar procedures (such as bone marrow transplants, reproductive procedures, or similar treatments using patients' own cells) have previously been treated as a part of medical practice and not subject to FDA regulations, other than meeting routine clinical laboratory standards. In the subsequent litigation, the FDA prevailed, the U.S. District Court (D.C.) finding that the cells were more than minimally manipulated. The company also maintained that the procedure was conducted entirely within one state (Colorado), and as such, it

[13] *Association for Molecular Pathology v. Myriad Genetics*, 569 U.S. 12–398.

should not be subject to the Federal Food, Drug, and Cosmetics Act (FDCA) which governs entities used in interstate commerce. The Court, however, found that because some components used in the procedure were shipped across state lines, the qualification of interstate commerce was met. The Court then went further, adding that the procedure interfered with the market for competing therapies because patients would travel to Colorado for the treatment, rather than having the same or alternative treatments in other states.[14]

The case will affect many cell-based therapies, especially autologous procedures which have previously been considered as an in-clinic physician procedure and therefore not subject to the FDCA, including potentially, many hematopoietic procedures which sort, process, and reinject cell components. The ruling about market effects is particularly subject to debate and raises questions about the scope of FDA authority. Interestingly, the case has been cast in some of the business literature as a patient rights issue, rather than as an argument between the stem cell industry and its regulators. Christopher Centeno, director of the Colorado clinic, expressed it this way: "We see this lawsuit as a twenty-first century civil rights issue that will define what control you have about the use of your own cells and tissue".[15]

The court cases, legislation, and regulatory challenges described above speak to the ongoing difficulty societies have in pinpointing what precisely is the thing that demands policy deliberation. What exactly is the problem to be solved? Is it the need to pin down the nature of the biological material itself, or is it the rights to access or change human bodily materials? Simply creating new or more defined categories is unlikely to fix this problem; new techniques will continue to disrupt whatever legal, social, or technical boundaries and terminologies we construct. Perhaps the problem is instead that existing institutional infrastructures are not adaptive to meet the needs of nascent science while maintaining public trust. Two things are clear: first, the way ethical problems are articulated will affect the range of solutions offered.

Second, it is clear that law and policy are not simply "responding to" novel science and technologies; rather, there are broader social and cultural undercurrents extending far beyond a single novel area of science, which cannot be ignored when considering governance of research practices.

---

[14] From the Memorandum opinion: "this admission [that process of culturing cells involves many steps] supports the conclusion that the biological characteristics of the cells change during the process employed by Defendants, resulting in more than minimal manipulation of the HCT/Ps originally extracted from the patient." Regarding the interstate commerce clause, the finding was that "one or more of its components is shipped via interstate commerce and because it substantially affects interstate commerce by depressing the market for FDA-approved out-of-state drugs by encouraging individuals to travel to Colorado to have the Regenexx Procedure performed instead. Therefore, because a component of the drug in this case is shipped through interstate commerce prior to its administration to the patient, the 'interstate commerce' requirement is also met." For the full opinion of *U.S. v Regenerative Sciences, LLC*, see https://ecf.dcd.uscourts.gov/cgi-bin/show_public_doc?2010cv1327-47.

[15] Many of the commentaries on the case came from general business sources, rather than stem cell science sources. See for example, Gottlieb and Klasmeier (2012) and Koleva (2012).

1.2.3 Translational Science Policy Initiatives

One of those currents is the intensification of efforts toward translational research and a policy environment favoring commercialization (cf Mason et al. 2012). A number of initiatives across the globe are promoting translational science to commercialize the products of science and bring new products to the market and the clinic (Etzkowitz et al. 2008; Fisher and Mauck 2013; Hellman et al. 2011; Rao 2009). A large number of RM-related companies are small firms established by academic entrepreneurs, and a large proportion of academic researchers have ongoing relationships with at least one for-profit company.

Translational policy and funding initiatives are included in programs from the Canadian Institute of Health Research, the UK's Research Governance Framework for Health and Social Care, Program 863 and Innovation 2020 in China, the Millenium Project in Japan, Germany's High-Tech Strategy 2020 (2010–2013), the America COMPETES Act, and other initiatives, all aiming to promote innovation for national economic gains.[16] Other initiatives, private organizations and advocacy groups, and public-private partnerships have been created with the express purpose of supporting translational work. The Genetics Policy Institute, the International Translational Regenerative Medicine Center, the Canadian Center for the Commercialization of Regenerative Medicine, the California Institute for Regenerative Medicine, regional groups such as the Pittsburgh Tissue Engineering Initiative, and others take on roles ranging from advocacy and community building to funding and project collaboration. Some of these groups are experimenting with novel organizational forms with licensing agreements for products arising from a pool of potential IP and guidelines adaptable for not-for-profit institutions and academic users.

Just like other medical technologies, a problem in getting RM products into clinical use is that while earlier stage research may be funded by conventional grants, in the period during which the research reaches a sufficient level to proceed, there is often little support or incentive to take the project on to the next stages, because of large costs and long development times before a product is ready for clinical trials (the so-called "valley of death"). Venture capital, larger pharmaceutical, or other firms may not be willing to carry the risk and the lack of guarantee of a high return on investment without evidence that the product will be successful. This has been especially true in RM where there is additional risk of uncertain regulatory, social, and legal environments. Recently, the National Center for the Advancement of Translational Science (NCATS) was created by the NIH to support activities in the zone between discovery research and human trials (http://www.ncats.nih.gov/).

---

[16] HR 2272, the "America Creating Opportunities to Meaningfully Promote Excellence in Technology, Education, and Science" (America COMPETES Act) is intended to strengthen research and education in the USA related to science, technology, engineering, and mathematics. The Act also establishes an Office of Innovation and Entrepreneurship to foster innovation and the commercialization of new technologies, in order to promote economic growth in the USA. It specifically contains provisions for open data sharing and responsible conduct of research, mandating RCR training for recipients of National Science Foundation grants (sec 7009–7012.) The Act was reauthorized in 2010 (HR 5116).

The center supports projects to develop alternative and more efficient preclinical tests (animal or computational analyses to determine risk and efficacy) to bridge the gap.

There are both opportunities and challenges for academic researchers following the commercialization lead. Academic-industry partnerships are increasingly important since industry has the capital and skills necessary to introduce products to markets that academic researchers often lack (Bayon et al. 2013). Commercialization puts money back into university coffers depleted by budget cuts. Also, as public funding for research shrinks, at least in the USA, new sources of funding will be needed to replace conventional NIH funding for both basic and translational research.

Others express concern that the resulting pressure to commercialize from such initiatives brings its own set of problems. Without a concomitant devotion to discovery research, there may be few innovations of sufficient quality to be developed commercially. There may also be financial and commitment conflicts of interest, as academic entrepreneurs become involved in marketing their inventions (see Chap. 6 for updates on rules regarding conflict of interest reporting). In a still-growing, young field like regenerative medicine, academics often take on policy advocacy or expert testimony roles, making it difficult to know when the effect of such activities might benefit a potential entrepreneurial interest (Munro 2003). There has been some concern that commercialization pressures would create inducements on academic researchers to pursue only certain kinds of research. There also might be a perceived breach of public trust if the expectations of public-private partnerships are not clearly spelled out; that is, a for-profit entity answering to the demands of private capital might define public goals quite differently than not-for-profit or academic researchers and thus there is a question about what constitutes a return on the taxpayers' investment in science. This raises questions about the appropriate relationship of academia, industry, and the state (Etzkowitz and Leydesdorff 2000; Herder and Brian 2008).

Conflicts of interests and motivations among academic scientists, technology transfer organizations (TTOs), and industrial partners may arise, as they frequently have differing expectations (Bubela and Caulfield 2010; George and Bock 2008; Siegel et al. 2007). Marketing, distributing, and supporting stem cell-based products requires more administrative and legal oversight than most other university technologies, which may discourage entrepreneurial interest by TTOs, yet academic scientists may not have the wherewithal to pursue such activities on their own. In a very telling survey of the North American tissue engineering community, all surveyed groups (academics, start-up companies, development stage companies, and established companies) perceived working with technology transfer offices to be even more of a hurdle than working with the FDA. For academics, this was number 8 of the most difficult barriers to commercialization, behind issues of obtaining funds (see Table 1.1 in Johnson et al. 2011).

A challenge for some scientists accustomed to conventional academic research is that entrepreneurial science also increases the expectations for impact measured in commercial or social terms. Beyond proof of principle or demonstrating that a therapeutic tool can be made using a stem cell technology, researchers must prove to funders that their project meets a strong societal or health need while providing a benefit, such as cost savings for health care, stimulus to the economy, or other social or economic impacts. These are analyses that stem cell researchers may not feel qualified to make and which may contribute to overpromising or "hype." Academic scientists are likely to have been trained to think primarily in terms of the research question itself, rather than downstream outcomes, whereas industrial researchers are accustomed to prioritizing market need and viability over product design or theory.

The issue comes up in considerations about education and training as well. It can be highly valuable to provide internships or other collaborative experiences for graduate students and postdocs in industry, as not all science students will end up in academic careers. Yet there are concerns for some about scholarly independence and vulnerability of students' and junior researchers' work, especially when potential for intellectual property or trade secrets is involved.

Finally, the pressure to conduct translational research may bring treatments into the clinic, but rushing to introduce experimental therapies into humans may have a high cost, not only for patients who may be harmed, but for the field as a whole. A recent editorial questioned the aggressive stance of the current Japanese government in proceeding with clinical trials on iPS cells. As part of its economic growth strategy, in June 2013, Prime Minister Shinzo Abe announced an innovation strategy which positions "substitution and supplement of body and organ functions" as one of its main pillars. The editorial suggested that economic goals were trumping clinical ones: "It is the role of the government to provide the right environment for research with a strong emphasis on safety and effectiveness. What it must not do is to be guided by economic motives."[17] It is not surprising that public concern and mistrust of government technology policies would be on the rise in Japan: in the wake of the Fukushima nuclear disaster, there has been a persistent perception of lack of community involvement in decisions that affect citizens' health and lives.

There is a need for public policy discussion more broadly to resolve potential conflicts between stated priorities of translational science and novel ways to achieve translation into public goods through a variety of public-private partnerships, for-profit organizations or other forms.

Chapters in this book address some of these tensions and how they are salient for regenerative medicine researchers. In the next chapter, they are put into global context.

1.3 Managing Knowledge Practices

1.3.1 Global Trends

The tensions between the need for discovery science and the need to commercialize products exist worldwide. As in other novel areas of innovation, regenerative medicine needs a major tangible success to continue to attract investments and talented

[17] Asahi Shimbun (editorial), found at http://ajw.asahi.com/article/views/AJ201306280044.

researchers and sustain public support. Yet there are competing social, ethical, political, and economic interests in individual societies about how best to prioritize health needs, research innovation, economic growth, and public benefits. What makes a research area such as regenerative medicine succeed or fail in the negotiations among various interests and priorities?

In light of the observation above that policy around novel science is contingently constituted, it is a mistake to oversimplify matters by claiming, as some commentators do, that research succeeds or fails only because of a lack of funding or because policies differ across locales. Such generalizations fail to acknowledge how science and science governance are deeply embedded in history, politics, and culture and the degree to which types of governance and scientific practices mutually constitute each other. In Chap. 2, Hogle, Palecek, Schaffer, and Zandstra draw on some of their collective, real-world observations to demonstrate that matters are far more complex than simple explanations allow.

Governments may make decisions for citizens based on a number of perceived social priorities: economic stimulation, historical experience with a technology or public reactions to it, consumer protection and safety, a pressing need to find treatments for illnesses or to reduce health care costs, and more. Public and private investors must weigh the perceived relative value of investing in types of technology and decide between competing needs—energy, hunger, health care, employment, and other pressing needs. Chapter 2 uses selected countries to demonstrate how governance over regenerative medicine funding and research policy must be understood within these contexts (see also Jasanoff 2005). In doing so, it sets the stage for more specific issues dealt with throughout the book.

1.3.2 Collaborations, Shared Resources, Open Science and Intellectual Property: Tensions or Opportunities?

In addition to tensions between policies promoting translational or discovery research, there are policies promoting commercialization (which by their nature encourage the sequestering of knowledge through intellectual property) and the national and international initiatives to foster "open science" (Rhoten 2007). On one hand, patents are important to start-ups and entrepreneurs not only to attract capital but to establish credibility. That is, there is value to the knowledge as much as the entity being patented. Patents are also seen as necessary to protect investments and the high costs that go into developing them. On the other, disallowing patents enables science to proceed unfettered by patents and allows greater transparency.

In Chap. 3, Bubela, Mishra, and Matthews detail some of these issues in relation to knowledge exchange across academic, industry, and state boundaries. The authors note that in the precompetitive space, an opportunity exists to share and leverage resources and knowledge. Labs may still be highly competitive with each other, racing to beat others to a discovery or to acquire funding and resources. However, in this phase, well-managed collaborations can result in collective learning to push the entire field forward, while being free from the transaction costs associated with licensing, patenting, and other restrictions to exchanging materials and information.

The authors describe various models of collaborative research, such as research commons, open-access models (such as the Sage Bionetworks, an open-source, collaborative computational platform), and public-private partnerships. Their insightful analysis is consistent with the stance of major international entities, such as the Organization for Economic Cooperation and Development (OECD) and the United Nations Educational, Scientific and Cultural Organization (UNESCO), both of which have guidelines recommending that open access should be an international norm. The NIH, ISSCR, and other governance bodies have endorsed open science models, arguing that open science offers the possibility of better science by avoiding duplication of results and learning from problems others have encountered, but more importantly, knowledge gained from research conducted with public funds should be shared and should be transparent. Despite such endorsements, some argue that the presumed norm of sharing and openness in science has been lost, and new models such as protected commons should be instituted (Winickoff et al. 2009).

Sharing of materials and information is not only a matter of intellectual property constraints; some researchers find it difficult to locate materials that may have been made by a member of a lab who has moved on, for example, and some labs simply will not share. The time and resources it takes to ship materials can take away from research, making this activity a lower priority. A core-service model could include consortia of participants across institutions and sectors, capitalizing on and leveraging particular kinds of expertise and materials at participating institutions. Industry-driven collaborative models with universities and professional societies are beginning to be employed at later stages, addressing downstream scale-up and manufacturing issues (Deans 2012).

There may be increasingly conflicting expectations between pressures to commercialize and pressures to have open-access models and transparency, particularly in novel, emerging science (Caulfield et al. 2012; Mathews et al. 2011). For example, as Pilar Ossorio notes in Chap. 5, the Human Genome Project has a prepublication data sharing policy, aiming to put information into the public domain, which may conflict with IP holders' motivations to withhold information.

Chapter 4 delves more deeply into intellectual property issues. In this chapter, Kevin Noonan provides a substantive history of important legal precedents with which to understand the context of patenting in stem cell research. As he demonstrates, the patenting of life forms has always been controversial, but there are additional moral components which have been viewed differently by various political and cultural systems. Noonan covers key cases, including the patenting of gene sequences (Myriad Genetics), metabolites (Prometheus Laboratories), and hESC cell lines (WARF). His insightful analysis is helpful in understanding how human-derived biological materials have come to be viewed by societies as scientific objects subject to patent law.

The European Union has directly addressed issues of patentability differently than the USA, for example. The political climate and ways of deliberating social issues particular to Europe set the stage for EU policy stances on science and technology, more generally, and specifically affected regenerative medicine through the banning of patents on products from human embryos.[18] While it sent shock waves throughout the stem cell industry, when viewed within historical and social contexts around biotechnology in the EU, the decision is not surprising. Unlike the USA, the European Union has viewed biotechnology more generally within a framework of human rights and dignity issues. This happened in the wake of social justice issues raised in the 1990s by TRIPS and WTO, Questions of access to medical products and therapies in resource-poor countries as well as the opportunity and right to develop science and technology that may benefit their own economy triggered policy responses differently than in North America. A resort to presumed sense of human dignity and morality has since been privileged in policy discussions. The European Patent Office had already been sensitized to issues related to patenting life forms after the 1998 University of Edinburgh patent approval in 1998 "Isolation, selection and propagation of animal transgenic stem cells" (the cloning of Dolly). In that case, the EU Parliament concluded that anything related to the destruction of an embryo (whether the cells themselves, processes around them, reporter genes, and other biological materials used) should not be allowed. The Parliament also subsequently opposed the granting of Myriad Genetics' patent for BRCA1 and two genes, again placing the decision about patenting human genetic material within a broader context of human dignity.

While Europe took an ethics-based stance on patenting using principles from Council of Europe for the protection of human rights and dignity, Asia and the USA have all recently updated laws to facilitate commercialization (see Chap. 2). In the USA, extensive patent reform took effect in March 2013. Named the Leahy-Smith "America Invents Act," it is designed to align US patent law with the rest of the world. One of the main changes is a switch from granting patent rights to the "first to invent" to the "first to file." Previously, scientists could produce proof that they were the first to make an invention, but scientists will need to be aware that they must file as quickly as possible, as the idea of being first with an idea will become irrelevant.

Much has been made of the inaccessibility of some hESC lines due to patenting restrictions, particularly in the early years, which created a cost barrier for smaller academic labs or start-up companies (Chapman 2009; Golden 2010; Torremans 2011). Noonan covers the history of embryonic stem cell patenting in detail. Some of these issues may become moot as some lines will go off patent soon. Also, for methods patents, patenting may play less of a role, as methods evolve so rapidly that older ones become quickly outdated, as in the case of iPS research. A related difficulty is that cell lines (both iPS and ES) and other RM-related technologies may have multiple patents associated with them. For example, someone may have genetically modified a line or added something which may have additional intellectual property components that may limit how the line may be used. This makes it difficult to identify to whom the IP should be attributed. Researchers can become

[18] Bergman and Graff (2007). See also Hogle et al. Chap. 2.

overwhelmed by having to chase down information to determine which lines may have embedded MTAs and multiple sources of IP and know under which conditions she or he can use the line. It is not clear to many researchers how these negotiations work. There are infrastructure and institutional barriers that add to the problem, such as private banks or cell suppliers which may not always provide complete and consistent information.

Just as important as patents, but less discussed, according to many researchers with whom I have spoken, is that of material transfer agreements (MTA), data use agreements (DUA), and memoranda of understanding (MOA) which may apply to cell lines as well as other necessary materials. These agreements are legal documents often containing restrictions of what can be done with the materials, yet many researchers informally share materials without considering the ramifications of what these agreements may restrict. Restrictions may have to do with what donors agreed to do (or did not agree to do), but there may be additional constraints imposed by the institution (university or commercial source).

Just as problematic are emerging issues around data sharing in an era of high-throughput biology. Technologies such as next-gen DNA sequencing and "omics" sciences have enabled the production of vast amounts of data and at lower costs. In North America and Europe, there are concerted efforts to keep this data in the public domain, consistent with the open science environment described above. This can be seen in the emergence of initiatives such as the database of Genotypes and Phenotypes (dbGaP), the European Genome-Phenome Archive (EGA), the i2b2 informatics framework (a public-private partnership funded by the NIH), and novel enterprises such as SAGE Bionetworks (Kesselheim and Karlawish 2012). Pilar Ossorio analyzes important legal, ethical, and policy implications of data sharing through such repositories in Chap. 5. There are two major policy issues: repositories require a structure for storing and accessing information from these large, mostly publicly funded databases in a way that encourages investigators to deposit and share their data, and there must be a system for controlling access and providing security to protect sensitive data. To incentivize data sharing, some funding agencies are requiring investigators to share their data by depositing it into an appropriate repository as a condition of funding, and journals are increasingly requiring or recommending authors to deposit data as a condition of publication.

Stem cell researchers will increasingly need to access raw data from these databases as well as the Stem Cell Discovery Engine and Stem Cell Omics Repository which has specific data on hES and iPS cell lines. While genetics researchers have known about the data sharing requirements for some time, some regenerative medicine researchers may be less aware of some of the access issues around repositories. As Ossorio notes, NIH genome-wide association study (GWAS) rules have been established for controlled access, that is, to control who has access to the data and under what conditions and to require investigators to use data security measures. Most challenging, however, is devising a way to access raw data without compromising sensitive personal information about the sources of the genomic information. Epigenome data within the same individual may be available in some repositories, which is useful for iPS research, and there may also be a desire to link genomic profiles with data from medical record data. Both activities could compromise anonymity and reveal sensitive information about individuals, yet it is becoming increasingly impossible to promise anonymity in light of the ease with which specimens can be re-identified (Gymrek et al. 2013; McGuire et al. 2011). Ossorio provides an important analysis of the difficulties inherent in attempting to protect sensitive genome-wide information while still allowing access and what to do if researchers request to do further studies beyond what the specimen donor consented to allow. New ways of considering informed consent are being considered in light of these emerging conditions (O'Doherty et al. 2011; see also Chap. 6).

1.4 Managing Responsible Research Practices

Part II turns to core research ethics issues for those planning discovery or clinical research in regenerative medicine-related areas, including recent changes to guidelines. For experienced researchers, students, and those new to the field, Chap. 6 provides an overview of crucial information to protect the integrity of the scientific endeavor and presents guideposts for how to recognize potential ethical concerns in their own or others' research. Chapters 7 and 8 deal with the protection of human subjects, particularly in "frontier research".

1.4.1 Research Integrity Calibrated for Regenerative Medicine Researchers

Chapter 6 (VonReyn, Das, and Hogle) lays out principles of research integrity and reviews the organizations involved in oversight authority. Investigators and their students who receive US federal funding (as well as those in many other countries) are required to have basic training in the responsible conduct of research (RCR). While there are many resources available with good information about the responsible conduct of science, many sources are necessarily general in nature. As such, some of the information and case studies used may be insufficient to meet the needs of contemporary interdisciplinary cell- and molecular-based science and engineering. For this reason, the authors give particular attention to circumstances that regenerative biology and engineering researchers may encounter. The principles and many of the illustrations in the chapter apply broadly to many forms of research, but the chapter also highlights additional issues particular to RM research, including those addressed by the National Academies of Science (NAS) and the International Society for Stem Cell Research (ISSCR). Additional topics include the use of chimeras for proof of principle, conflicts of interest when the researcher may also be the practitioner responsible for a potential biospecimen donor's health care, and problems that might arise when using specialized equipment and techniques used in RM.

The authors argue that contemporary research integrity dilemmas may not fit neatly into discrete "checkboxes" designated by oversight authorities. RCR training materials may be prepared by those knowledgeable about the formal rules of oversight but may not be up to speed on contemporary dilemmas in everyday scientific practices. Furthermore, investigators are frequently confronted by situations that lie in gray areas where it is unclear what counts as misconduct. Some of these situations may arise in the course of using new equipment technologies or in the uncertainty of how to interpret and represent findings. Additionally, much of the information published about cases of misconduct focuses on the failures of individual investigators (either through negligence, intention, or just not knowing what to do or not do)—the so-called "bad apple" theory of misconduct. Yet there may also be system failures. For example, a system of peer review exists for publication and funding, but there may be conflicts of interest or lack of expertise that fails to identify potential research integrity problems. Stem Cell Review and Oversight Committees and Institutional Review Boards exist, but there is little consistency among institutions (including IRBs at academic or research institutions and for-profit organizations) in the way protocols are reviewed and the way ethical dilemmas are. The solution may not be simply adding more institutional rules or additional oversight bodies; in fact, this may just make people more complacent, assuming that everything is taken care of, unless a culture of individual, organizational, and societal responsibility is instilled (Hogle 2009). The chapter encourages readers to be discriminating and thoughtful about their own practice as well as the system of research oversight overall in hopes that investigators will engage with policymakers about best practices for complex, contemporary, interdisciplinary science.

Sections 6.3.1 and 6.3.2 address key issues in human and animal research subject protection, including upcoming changes in guidelines and providing background to help understand why the changes. The authors note that conventional bioethics tends to privilege Western notions of autonomy and focus almost entirely on the rights of individuals, but ultimately, research is fundamentally a social enterprise. As such, the relationship with subjects—without whom research could not proceed—must not be abused (London et al. 2010).

1.4.2 Protecting Research Participants

The welfare of human and animal subjects in research is essential to maintaining trust in scientific endeavors and is dealt with in the final two chapters. Trust among donors, clinicians, and researchers is essential to protect the integrity of scientific research, and properly handling consent to obtain and use an individual's tissue or their medical information is critical to maintaining that trust (Haimes and Taylor 2011). In Chap. 7, David Resnik describes how best to protect the interests of individuals who donate biological materials used in RM research.

Asking a couple to donate their unused embryos is particularly ethically sensitive, so there must be consistency and care in both the content of the consent and the timing of the request. Recognizing this, the revised 2009 NIH guidelines for hESC research prioritized information about the informed consent process (including the manner in which information was taken and what was promised to donors) as a criterion on which eligibility for federal funding would be determined. This change in policy had direct effects on research: cell lines derived without the consent requirements as set forth in the new guidelines were not included on the new NIH registry, even though some of the embryos were donated long before the new guidelines came into effect, and the various private clinics and research institutes where embryos had been donated each had their own institution's consent forms (see also Chap. 6, Sect. 6.3 for a discussion of the ramifications). Additionally, dozens of hESC lines with mutations for specific diseases, which might have been extremely useful for a range of disease studies, were rejected from eligibility for study using NIH funds, because the language in the donor consent forms did not match the new specifications (Wadman 2010b).

Consent from adults donating their own biological materials also generates questions about what to include in consent agreements. How detailed should consent forms be, and is it necessary to list all possible eventual research uses? What if some experiments are not possible or even imagined at the time of asking for the donation? Potential research uses could be stated extremely broadly, but this opens the door to uses for which the donor would not consent. If, on the other hand, consent is specified for a particular use (such as diabetes research), and if a cell line created from that donated material is found to be extremely useful for another purpose (such as cardiovascular research), can it still be used? This scenario has already occurred, and when interpreted in the strictest sense of the new guidelines, such new uses would not be allowed. Another concern that has arisen is the issue of whether or not potential donors should be told that products from their materials may be commercialized. Legal history has shown that the question of who might have property rights in human biological materials (including one's own tissue) is as fraught an issue as that of what kinds of "natural" materials may be patented.[19] More empirical data is needed to find out how donors might respond to this information and to what extent it would affect their participation in research.

Resnik also discusses procedures intended to maintain confidentiality (the guarantee that researchers will not share personal information about donors) and anonymity (the guarantee that personal information about the donor will not be identifiable during the course of research).

Elaborate de-identification methods have been created to assure anonymity, but as described above, it is almost impossible to guarantee. Re-identification algorithms can trace information back the donor's identity, with enormous implications for privacy and possible stultifying effects on voluntary donation. Yet researchers may need to know medical histories of donors or may need to obtain further genetic information, either to conduct research on specific diseases or simply to protect

[19] cf *Moore v Regents of the University of California* (51 Cal. 3d 120; 271 Cal. Rptr. 146; 793 P.2d 479. An excellent discussion of the legal and ethical issues of ownership in human biomaterials is found in (Charo 2006).

against problems that may arise with an immortalize cell line. There is potential for significant effects on research in the near future, as discussed at the end of this introductory chapter.

Clinical trials are already underway for some stem cell and tissue engineering therapies. Some of these, particularly those involving autologous infusion of cells and cell products, are similar enough to clinical procedures (such as bone marrow transplantation or autologous cell infusions) that existing practices may suffice for the design and execution of trial protocols. Based on this assumption, it appears that some clinicians are trying cell-based treatments on their own patients without conducting a formal clinical trial. However, most of regenerative medicine consists of experimental treatments which have not yet been tried in humans. As such, there is a need to be prepared for potential, unanticipated risks. In Chap. 8, Nancy King discusses critical issues in planning for human trials and outlines ethical protections unique to regenerative medicine. In particular, she argues, it is important to choose appropriate animal models and to know when it is safe to move from preclinical testing to human experimentation. Recent findings confirm that mouse models do not necessarily recapitulate human responses and may not predict how a therapy will work in humans (Seok et al. 2013). While not surprising, the finding indicates that more work needs to be done to establish safety before testing RM products in humans, and more information is needed about alternative preclinical testing modalities.

Cell-based therapies will be particularly challenging, in part because of unique safety issues related to pluripotency, but also because efficacy will be very difficult to prove plus endpoints may be difficult to define. Several countries are disallowing the use of surrogate endpoints, looking instead for firm clinical endpoints—but what precisely should these be for cell-based treatments on a variety of tissue types and functions? When cells or bioactive materials are implanted therapeutically, it may take a long time for tissue remodeling to occur, and homologous function may be difficult to define or determine. Also, the "gold standard" randomized trial may not be possible, particularly with entities such as surgically implanted tissue-engineered products which could not be blinded or easily randomized. This raises a number of questions about trial design and what data is needed before a trial can begin: At what points—and for how long—must patients be followed post trial?

King includes two other forms of risk: first, subjects may not clearly understand the unique features of a cell-based intervention, including the distinction between trials (research) and a treatment, but also how such a trial differs from a drug trial. Second, there is some risk that information circulated about early trials may be misinterpreted in the public sphere, creating unduly high expectations that successful treatments or "cures" are around the corner. This concern becomes particularly acute with the rise in the number of unproven therapies being offered around the world (see Chap. 2, Sect. 2.1.3 on stem cell tourism).

It is important for researchers and clinicians to know what is required by the guidelines established by their national or local governance bodies, so as not to lose opportunities to gather valuable information as well as not to lose the trust of participants. At the same time, informed consent has become somewhat fetishized; that is, there is a tendency of many bioethicists and policymakers to focus on matters of process and rule-making structures rather than matters of content. Assuming that simply having an informed consent process in place will take care of all ethical concerns may provide a false sense that ethical duties have been met, and focusing attention on the consent process alone displaces attention away from other salient concerns, which often receive less attention. Policymakers and ethicists need to stay abreast of both the scientific and institutional changes happening in regenerative medicine as they occur, without relying on research subjects and tissue donors' consent process to do all the work of managing expectations associated with medical innovation.

1.5 Emerging Issues

There are several themes recurring throughout the chapters that speak to broader social issues. These include tensions between open science and private intellectual property, new multi-sector arrangements that may reorganize academia-industry-state relations, changes in the meaning of ethical "protections" with new challenges to data privacy, and the conflicts inherent in knowing when and how best to proceed with novel science and technology in light of the tensions between unknown (and sometimes unknowable) outcomes. These issues and more will require careful consideration by scientists, policy makers, ethicists, and those involved in commercializing regenerative medicine products and planning the next generation of scientific and technological innovations.

New techniques will undoubtedly raise fresh dilemmas about what is ethically permissible, and many will fall into the gray areas for which guidelines do not exist. Just as important as innovations in the science are innovations in institutional infrastructures and governmental priorities. One example bearing close observation over the next few years is the recent drive by the NAS and NIH to transform medicine into so-called "precision medicine" (NAS 2011) and similar initiatives in the EU to create large-scale data infrastructures linking medical and potentially, civil databases.

The Precision Medicine Report calls for an interconnected "knowledge network"—a sort of information commons—in which data from biorepositories, electronic medical records and clinical health records, "omics" data, and potentially, information gleaned from data analytics on social media can be linked and analyzed for associations. Behavioral, social, and environmental data (the "exposome"—a measure of lifelong environmental and lifestyle exposures) would be layered to pinpoint more precisely the causation of disease and possibly identify prevention or intervention measures. Access to such rich data may be extraordinarily helpful to researchers trying to link disease phenotypes to genotypes and predict responses to therapeutic agents. One promised possibility of personalized medicine using regenerative medicine is the capability to model diseases using patient iPS cells, then connecting information from such linked databases to in vitro findings (Park et al. 2008). In this case it would likely be necessary to have patients and their families more directly involved in the research, by contributing biospecimens and family medical histories and possibly by providing information about symptoms not captured in medical records. This will be particularly necessary where disorders are more complex than single gene mutations; factors involved could be related to various epigenetic influences, for which it would be helpful to have environmental and family histories. Because large numbers of individuals will need to be biopsied and their cells stored in biobanks to test hypotheses about phenotypes, a number of public and private initiatives have proliferated to recruit patients into voluntarily providing information.

Much of the patient's private information, however, may be accessible without asking directly, either from stored records or information and samples that could be collected in the course of routine clinical encounters. The NAS report calls for easing of the existing guidelines requiring specific patient-informed consent for such research, such that it may not be necessary to request consent for additional uses beyond those agreed upon in the original consent. This is consistent with the proposed guideline changes discussed in Chap. 6 regarding human subjects protections. At the date of this writing, these guidelines have yet to be decided. This situation has raised concern for many regarding the nature of consent, what counts as "private information," and who has the right to use individuals' health information. It may also pit rights and entitlements of individuals against the need to solve complex problems of disease prevention and treatment, a long-standing tension in public health.

As Ossorio shows in her chapter, a large number of the US population can already be easily uniquely identified using a combination of simple data (birth date, zip code, voter records) easily available from public sources. This begs the larger question of what privacy means in the post-genomic, post-internet era. Can we really guarantee anonymity to biospecimen donors and human research subjects, as we have for decades? The protection of privacy has long been a central tension in the ethics of public health, from the need to report and surveil infectious diseases pitted against concerns about stigma to potential effects on employment or insurance discrimination created by the disclosure of genetic susceptibilities to the simple act of respecting a hospital patient's medical situation. Do individuals *care* whether their information is used in this way, or not? (Hoeyer et al. 2005). Is the problem really about control over information rather than privacy, or who has the right to say how information is used? Empirical data needs to be collected before instituting rules to know how best to proceed.

In any case, the roles and relationships of patients, clinicians, and researchers will shift as a result of such major institutional and informational infrastructure shifts, particularly in personalized medicine using RM. The modeling of disease that might arise (in vitro, in vivo, and in silico) may also change not only disease theories but the way we organize, fund, and otherwise deal with clinical health delivery, public health, and governance (including research ethics).

These issues do not exist in a vacuum; rather, they lie within broader social, political, and economic shifts taking place in contemporary societies. As this book was about to go to press, alarms were being raised worldwide about governmental uses and potential abuses of private information in the wake of leaked information indicating that the US government had accessed personal records of thousands of citizens. How might legal and social concepts of privacy be changing in light of competing social values of national security and protection of personal information and liberties? Bringing to light activities that have not been publicly visible may likely affect issues in other social domains, such as medicine.

This example is one among many that demonstrates how science and science policy interact and shape each other within historical and political moments and how policy actions may be triggered (or not) by news events, legal challenges, powerful political groups, or other influences. It is in these interactions that rights and entitlements are defined, and boundaries of jurisdictions are settled. The boundaries can be fluid, as examples in this book and the policy updates at the beginning of this chapter show. It is my hope that readers will critically engage with the information provided here, use what is provided to prepare for new challenges, and that scientists from academia and industry, ethicists, policymakers, and social scientists can learn from and engage with each other.

References

Asahi Shimbun (2013, June 29). Too much of a rush for clinical trials with iPS cells (editorial). Retrieved June 30, 2013, from http://ajw.asahi.com/article/views/AJ201306280044.

Bayon, Y., Ellis, S., Vertes, A., Ahmed, A., Coury, A. J., Campion, C., et al. (2013). Commercialization of regenerative products—The academic/industry partnership—Society World Congress: Summary report: Vienna, Austria/September 5–8, 2012. *Tissue Engineering. Part B, Reviews*. doi:10.1089/ten.TEB.2012.0683.

Bergman, K., & Graff, G. D. (2007). The global stem cell patent landscape: Implications for efficient technology transfer and commercial development. *Nature Biotechnology, 25*(4), 419–424.

Brock, D. (2010). Creating embryos for use in stem cell research. *The Journal of Law, Medicine & Ethics, 38*(2), 229–237.

Bubela, T. M., & Caulfield, T. (2010). Role and reality: Technology transfer at Canadian universities. *Trends in Biotechnology, 28*(9), 447–451.

Caulfield, T., Harmon, S., & Yann, J. (2012). Open science versus commercialization: A modern research conflict? *Genome Medicine, 4*(2), 17.

Chapman, A. R. (2009). The ethics of patenting human embryonic stem cells. *Kennedy Institute of Ethics Journal, 19*(3), 261–288.

Charo, R. A. (2001). Every cell is sacred: Logical consequences of the argument from potential in the age of cloning. In P. Lauritzen (Ed.), *Cloning and the future of human embryo research* (pp. 82–89). Oxford: Oxford University Press.

Charo, R. A. (2006). Body of research—Ownership and use of human tissue. *The New England Journal of Medicine, 355*(15), 1517–1519.

Cohen, C. (2007). *Renewing the stuff of life: Stem cells, ethics and public policy*. New York: Oxford University Press.

Collins, L. R., & Crockin, S. L. (2012). Fighting "personhood" initiatives in the United States. *Reproductive Biomedicine Online, 24*(7), 689–691. doi:10.1016/j.rbmo.2012.04.002.

Condic, M. L., & Rao, M. (2008). Regulatory issues for personalized pluripotent cells. *Stem Cells, 26*(11), 2753–2758.

de Vries, R., Oerlemans, A., Trommelmans, L., Dierickx, K., & Gordijn, B. (2008). Ethical aspects of tissue engineering: A review. *Tissue Engineering. Part B, 14*(4), 367–375.

Deans, R. (2012). Regulation, manufacturing and building industry consensus. *Regenerative Medicine, 7*(6 Suppl), 78–81.

DiSilvestro, R. (2006). Not every cell is sacred: A reply to Charo. *Bioethics, 20*(3), 146–157.

Downs, K. (2008). Embryological origins of the human individual. *DNA and Cell Biology, 27*(1), 3–7. doi:10.1089=dna.2007.0698.

Dubow, S. (2011). *Ourselves unborn: A history of the fetus in modern America*. New York: Oxford University Press.

Etzkowitz, H., & Leydesdorff, L. (2000). The dynamics of innovation: From national systems and 'mode 2' to a triple helix of university-industry-government relations. *Research Policy, 29*(2), 109–123.

Etzkowitz, H., Ranga, M., Benner, M., Guaranys, L., Maculan, A. M., & Kneller, R. (2008). Pathways to the entrepreneurial university: Towards a global convergence. *Science and Public Policy, 35*(9), 681–695.

Faulkner, A. (2009). Regulatory policy as innovation: Constructing rules of engagement for a technological zone of tissue engineering in the European Union. *Research Policy, 38*(4), 637–646.

Fink, D. W. (2009). FDA regulation of stem cell-based products. *Science, 324*(5935), 1662–1663.

Fisher, M. B., & Mauck, R. L. (2013). Tissue engineering and regenerative medicine: Recent innovations and the transition to translation. *Tissue Engineering. Part B, Reviews, 19*(1), 1–13.

George, G., & Bock, A. J. (2008). *Inventing entrepreneurs: Technology innovators and their entrepreneurial journey*. Upper Saddle River, NJ: Pearson Prentice Hall.

Golden, J. M. (2010). WARF's stem cell patents and tensions between public and private sector approaches to research. *The Journal of Law, Medicine & Ethics, 38*(2), 314–331.

Gottlieb, S., & Klasmeier, C. (2012, August 7). The FDA wants to regulate your cells. *Wall Street Journal*. Retrieved July 1, 2013, from http://online.wsj.com/article/SB1000087239639044440 5804577558992030043820.html.

Gottweis, H., Salter, B., & Waldby, C. (2009). *The global politics of human embryonic stem cell science: Regenerative medicine in transition*. Basingstoke: Palgrave Macmillan.

Gymrek, M., McGuire, A., Golan, D., Halpern, E., & Erlich, Y. (2013). Identifying personal genomes by surname inference. *Science, 339*(6117), 321–324. doi:10.1126/science.1229566.

Haimes, E., & Taylor, K. (2011). Researching the relationships between tissue providers, clinicians and stem cell scientists. *Cell Stem Cell, 8*(6), 613–615.

Hellman, K. B., Johnson, P. C., Bertram, T. A., & Tawil, B. (2011). Challenges in tissue engineering and regenerative medicine product commercialization: Building an industry. *Tissue Engineering. Part A, 17*(1–2), 1–3. doi:10.1089/ten.TEA.2010.0436.

Herder, M., & Brian, J. (2008). Canada's stem cell corporation: Aggregate concerns and the question of public trust. *Journal of Business Ethics, 77*(1), 73–84.

Hoeyer, K., Olofsson, B.-O., Mjörndal, T., & Lynöe, N. (2005). The ethics of research using biobanks: Reason to question the importance attributed to informed consent. *Archives of Internal Medicine, 165*(1), 97–100. doi:10.1001/archinte.165.1.97.

Hogle, L. F. (2009). Science, ethics and the "problems" of governing nanotechnologies. *Journal of Law, Medicine & Ethics, 37*(4), 749–758.

Holland, S., Lebacqz, K., & Zoloth, L. (Eds.). (2001). *The human embryonic stem cell debate: Science, ethics, and public policy*. Cambridge, MA: MIT Press.

Jasanoff, S. (2004). *States of knowledge: The co-production of science and social order*. London: Routledge.

Jasanoff, S. (2005). *Designs on nature: Science and democracy in Europe and the United States*. Princeton, NJ: Princeton University Press.

Johnson, P. C., Bertram, T. A., Tawil, B., & Hellman, K. B. (2011). Hurdles in tissue engineering/regenerative medicine product commercialization: A survey of North American academia and industry. *Tissue Engineering. Part A, 17*(1–2), 5–15. doi:10.1089/ten.tea.2010.0411.

Kesselheim, A. S., & Karlawish, J. (2012). Biomarkers unbound—The Supreme Court's ruling on diagnostic-test patents. *The New England Journal of Medicine, 366*(25), 2338–2340. doi:10.1056/nejmp1204164.

Koleva, G. (2012, February 10). Stem cells and the lawsuit that could change our medical future. *Forbes*. Retrieved July 1, 2013 from http://www.forbes.com/sites/gerganakoleva/2012/02/10/stem-cells-and-the-lawsuit-that-may-shape-our-medical-future/.

Liptak, A. (2013, June 13). Justices, 9–0, bar patenting human genes. *The New York Times*. Retrieved June 13, 2013, from http://www.nytimes.com/2013/06/14/us/supreme-court-rules-human-genes-may-not-be-patented.html?ref=myriadgeneticsinc.

London, A. J., Kimmelman, J., & Emborg, M. E. (2010). Beyond access vs. protection in trials of innovative therapies. *Science, 328*(5980), 829–830.

Maienschein, J. (2002). What's in a name? Embryos, clones and stem cells. *The American Journal of Bioethics, 2*(1), 12–19.

Marshall, E. (2013). Supreme Court rules out patents on "natural" genes. *Science, 340*(6139), 1387–1388.

Mason, C., McCall, M. J., Culme-Seymour, E. J., Suthasan, S., Edwards-Parton, S., Bonfiglio, G. A., et al. (2012). The global cell therapy industry continues to rise during the second and third quarters of 2012. *Cell Stem Cell, 11*(6), 735–739. doi:10.1016/j.stem.2012.11.013.

Mathews, D., Graff, G., Saha, K., & Winickoff, D. (2011). Access to stem cells and data: Persons, property rights, and scientific progress. *Science, 331*(6018), 725–727.

McGuire, A. L., Oliver, J. M., Slashinski, M. J., Graves, J. L., Wang, T., Kelly, P. A., et al. (2011). To share or not to share: A randomized trial of consent for data sharing in genome research. *Genetics in Medicine, 13*(11), 948–955. doi:10.1097/GIM.0b013e3182227589.

Morgan, L. M. (2009). *Icons of life: A cultural history of human embryos*. Berkeley, CA: University of California Press.

Munro, N. (2003). Academics' ties to business muddy disclosure decisions. *The Scientist, 17*(8), 54–55.

National Academies of Science. (2011). Toward precision medicine: Building a knowledge network for biomedical research and a new taxonomy of disease. Retrieved October 2012, from http://www.nap.edu/catalog.php?record_id=13284.

National Institutes of Health. (1994). *Report of the human embryo research panel*. Bethesda, MD: National Institutes of Health.

O'Doherty, K. C., Burgess, M. M., Edwards, K., Gallagher, R. P., Hawkins, A. K., Kaye, J., et al. (2011). From consent to institutions: Designing adaptive governance for genomic biobanks. *Social Science & Medicine, 73*(3), 367–374. doi:10.1016/j.socscimed.2011.05.046.

Park, I.-H., Arora, N., Huo, H., Maherali, N., Ahfeldt, T., Shimamura, A., et al. (2008). Disease-specific induced pluripotent stem cells. *Cell, 134*(5), 877–886.

Rao, M. S. (2009). Funding translational work in cell-based therapy. *Cell Stem Cell, 9*(1), 7–10. doi:10.1016/j.stem.2011.06.009.

Rhoten, D. (2007). The frontiers of intellectual property: Expanded protection versus new models of open science. *Annual Review of Law and Social Science, 3*(1), 345–373. doi:10.1146/annurev.lawsocsci.3.081806.112900.

Seok, J., Warren, H. S., Cuenca, A. G., Mindrinos, M. N., Baker, H. V., Xu, W., et al. (2013). Genomic responses in mouse models poorly mimic human inflammatory diseases. *Proceedings of the National Academy of Sciences of the United States of America, 110*(9), 3507–3512. doi:10.1073/pnas.1222878110.

Siegel, D. S., Veugelers, R., & Wright, M. (2007). Technology transfer offices and commercialization of university intellectual property: Performance and policy implications. *Oxford Review of Economic Policy, 23*(4), 640–660.

Singer, P., & Kuhse, H. (1986). The ethics of embryo research. *The Journal of Law, Medicine & Ethics, 14*(3–4), 133–138.

Torremans, L. (2011). Legal problems raised by patents on human stem cell-based interventions. In K. Hug & G. Hermerén (Eds.), *Translational stem cell research: Issues beyond the debate on moral status of the human embryo* (pp. 287–307). New York: Humana Press.

Wadman, M. (2010a). US stem-cell chaos felt abroad. *Nature, 467*(7312), 138–139. doi:10.1038/467138a.

Wadman, M. (2010b). Diseased cells fail to win approval. *Nature, 465*, 852.

Winickoff, D., Saha, K., & Graff, G. (2009). Opening stem cell research and development: A policy proposal for the management of data, intellectual property and ethics. *Yale Journal of Health Policy, Law, and Ethics, 9*(1), 52–127.

Chapter 2
Characterizing International Stem Cell Research Niches

Linda F. Hogle, Sean P. Palecek, David Schaffer, and Peter W. Zandstra

2.1 Introduction

This chapter assesses global research climates for stem cell-related research, analyzing infrastructure, governance, and funding patterns within regions and their interactions with social, historical, and political influences. Our findings derive primarily from site visits, interviews, and experience working with stem cell researchers in multiple research institutions in North America, Europe, and Asia. Authors Palecek, Schaffer, and Zandstra participated in a global assessment of stem cell engineering to identify emerging innovations, identify opportunities and barriers in the field, and provide information for funding agencies for the future (Nerem et al. 2013). Hogle has interviewed stem cell scientists and engineers in North America and Europe for more than a decade. We add to these firsthand observations some contextual understanding of why and how stem cell research more broadly has emerged in the way it has in different regions. In this chapter, we compare the contexts of global research environments. The chapter is not meant to be a comprehensive global

L.F. Hogle (✉)
Department of Medical History & Bioethics, University of Wisconsin-Madison
1300 University Ave 1135 Medical Sciences Building Madison, WI 53706, USA
e-mail: lfhogle@wisc.edu

S.P. Palecek
Department of Chemical & Biological Engineering, University of Wisconsin - Madison,
1415 Engineering Drive, Madison, WI 53706, USA

D. Schaffer
Department of Chemical and Biomolecular Engineering, University of California,
Berkeley 274 Stanley Hall, Berkeley, CA 94720-3220, USA

P.W. Zandstra
Canada Research Chair in Stem Cell Bioengineering, Terrence Donnelly Centre for Cellular and Biomolecular Research Donnelly (CCBR) Building, 160 College Street, Office # 1116,
University of Toronto,
Ontario M5S 3E1, USA

L.F. Hogle (ed.), *Regenerative Medicine Ethics: Governing Research and Knowledge Practices*, DOI 10.1007/978-1-4614-9062-3_2,

analysis, which would take several volumes. Also, available published data and analyses are uneven; information on current governance, capital and financing, or stem cell training is more readily accessible for some regions than others. Instead, we have selected countries with which we are more familiar and which have salient features to discuss to illustrate our points. We begin with an overview of general trends we have observed, followed by activities in regional niches. We do not pretend to have sufficient empirical data to provide precise prescriptions; rather, we offer the benefit of our individual and collective observations through study of and collaboration with labs across these regions.

While many existing analyses of stem cell activities focus on cell therapy applications, we include stem cell-related diagnostics, disease modeling, drug discovery, and more, all of which require expertise from various engineering and computational sciences as well as biological sciences. More than simply identifying locales and describing existing research activity, our observations help to illuminate why and how research has been shaped as it has in various regions. This insight will provide a window to understanding where opportunities exist for improvements in science policy and, more specifically, research program planning for regenerative biosciences.

Existing descriptions of the state of the field often assume that regenerative medicine has developed where and how it has because of ethical or regulatory environments, which either constrain or support the research. Certainly policy that restricts the use of particular procedures or cell types limits access to materials and blunts knowledge, and there is a possibility of a brain drain as researchers move to locales more favorable to stem cell research. There is some evidence suggesting that researchers may shift their research as a result of the sheer uncertainty. Other feasible assumptions are that particular locales dominate because of powerful key scientists or political advocates who drive the research or because funding is more easily available. While there is no doubt that these elements affect the direction and tempo of any research, we argue that these explanations alone are too simple: there are far more complex interweavings of political, historical, economic, moral, and technical specificities that shape regenerative medicine research differently within locales. This is not to say that stem cell research microenvironments exist in isolation: regional and local-level conditions interact with international trends in much the same way as stem cell niches in the body affect and are affected by their interactions with other bodily systems.

In some of the countries we include in our discussion, research flourishes in spite of a lack of policy or regulatory clarity or with funding flows that are less than optimal for the particular needs of stem cell research. In other areas, it has failed to get traction in ways one might have predicted. Taking a fresh look, without taken-for-granted assumptions, we can ask different questions. Rather than simply asking where the favorable or unfavorable policies are, we can ask broadly what are the most important situational components that appear to help or hinder research? When we understand the contexts in which regenerative medicine science and policies are shaped, it is possible to identify needs and promote policies to enable researchers to capitalize on opportunities: where are natural avenues for collaborations, and how have collaborations been supported (or not) in various environments? What are identifiable skills and resources needs for the longer term as well as immediate needs? As we learned, there is insufficient data on some regions to make comparisons evenly. What information is needed to provide insight into what governance forms do or do not work well? We hope that our observations about the experiences in various countries will stimulate fresh strategic thinking about how to develop policies to transcend differences in ways that will meet real global health needs while supporting both translational and discovery research.

2.2 Global Patterns of Activity

Our review shows that where regenerative sciences are emerging, there are diverse rationales for pursuing regenerative medicine, various ways of organizing and funding research, and differing degrees to which collaborations are being formed (within or across national borders). Countries may consider investing in innovative science (and regenerative medicine in particular) for economic growth, a way to grow innovation processes or science infrastructures or a way to serve the country's health needs. Politics and science thus are co-produced. Development of a new area requires the political will to do so, which in turn affects regulatory policies, legislation, and funding.[1]

2.2.1 *Investments in Global Knowledge Economies*

Funding is central to the development of new scientific endeavors. However, it is not only the level but the type of financial arrangements that matter. Funding in Singapore, China, Japan, and the state of California, among other locales, has been a part of strategic government-sponsored initiatives. In other countries, there may not be an explicit national strategy or coordinated effort, but a mix of public and private funding is available.

The countries which have made stem cell research a central part of a national science strategy are not necessarily those with the kind of existing industries capable of taking up product development once innovations leave the lab (Salter 2009a). Countries still in development and those which had devastated economies after World War II focused on building specific industries in the second half of the twentieth century that would aid in rebuilding and enable their participation in global markets. These were largely manufacturing and mass production technologies.

[1] It is also important to remember that bounded entities such as "the state" may be friction with supranational entities such as the European Union, the World Trade Organization, and other transnational entities attempting to enforce global harmonization of policies, definitions, and practices (Jasanoff 2004).

By the end of the century, there was a recognition that expertise was equally important as a national resources as production capacity. The transition to "knowledge economies" required specialized labor competencies, and the global focus shifted to communications and life sciences (Etzkowitz and Leydesdorff 2000). However, the life sciences are less developed in many of these countries, and regenerative medicine is uncertain with high initial investment costs. Also, developing nations, particularly those with central economies, have little experience or institutional structure with which to proceed with scientific or clinical innovation, especially when it is market based. In economies that are restructuring, the capacity to refocus national commitment comes with a concomitant need to restructure laws and social policies, in order to be globally competitive. As we have seen with several countries, this has been a struggle when approaching multifaceted life sciences sectors, especially those such as stem cell research which have faced public controversy.

We note that translational approaches dominate in almost every locale pursuing regenerative medicine research. While "translational medicine" is most simply defined as getting ideas from "bench to bedside," indicating the application of research for clinical purposes, the term is often taken to mean "commercialization," since product development, production, marketing and pricing are involved in getting products into circulation. This has implications both for academic-industry relations and for research funding. Conventional funding mechanisms are designed to support individual labs in discovery-level, disciplinary-based research. The kind of interdisciplinary, often pragmatic work that is necessary for bench-to-bedside work has not historically been rewarded by grant mechanisms or through merit systems in the university. However, some new funding initiatives are beginning to support translational work, such as EU Framework initiatives built around a particular problem, encouraging multi-investigator projects and potentially aimed at clinical translation. The National Center for the Advancement of Translational Science (NCATS) within the NIH was established in 2011 to support activities to "reduce, remove or bypass bottlenecks in the development of new treatments and tests that will ultimately improve human health" (http://www.ncats.nih.gov/). Translation-related initiatives and centers include the Berlin-Brandenburg Center for Regenerative Therapies, Canadian Center for the Commercialization of Regenerative Medicine, Fraunhofer IZI, McGowan Institute for Regenerative Medicine, and the California Institute for Regenerative Medicine (CIRM), among others.

The participation of strong private funds and venture capital in novel areas of science depends largely on whether decentralized, private investment is encouraged or discouraged by financial and scientific regulatory institutions (Bruton et al. 2005; Salter and Salter 2010). As Salter points out, the effect of investments in innovation relies on a country's environment for intellectual property rights, tax laws or incentives, and managerial freedom to exploit the value of an innovation as much as on the nature of the innovation itself (Salter 2009a). These factors help potential investors weigh risk against potential gains. In this sense, the environment has been most favorable in the USA, which had 76 % of the global total of venture capital funds in 2011 (Burrill 2012). Nevertheless, there is still a significant problem of the "valley of death" for start-up businesses; that is, the trough between obtaining angel or venture capital before the product is able to bring in revenue. For regenerative medicine

(RM) firms, there is also often a gap between initial funding for early phase clinical trials and the enormous investment needed for scaling up for phase II trials. Especially since the financial crisis in 2008, potential investors have been more risk averse, and pharmaceutical and biotech companies have been less inclined to invest in the highly risky market for clinical stem cell products, particularly in locales with more formidable or unsettled regulatory climates (Brindley et al. 2011; Pagnol et al. 2009; Rao 2009). This situation may be changing as the industry picture improves, but it is too early to tell (Mason et al. 2012; see also Martin et al. 2006).

For highly specific, high-tech enterprises such as regenerative medicine, intellectual property (IP) is a valuable resource. For some high-tech fields, patents may be worth less because the technology goes out of date quickly, but patent protection is seen to be paramount because of the high costs and long development times. Patents have become central to the uptake and growth of innovative science, but firms (especially new start-ups) must also have the kind of knowledge and expertise needed to attract talent, collaborators, and investors in the global market: "For capitalization of a new knowledge market to occur, investors need to be reassured that the *value of the knowledge*, as opposed to the value of the eventual product, is in the hands of the company concerned" (Salter 2009b, p.411, emphasis added). Knowledge includes not only scientific expertise but also expertise in all the processes needed to scale up, distribute, and get products into the hands of users. A country with a history of a strong centralized economy transitioning to a market-based economy is unlikely to have a strong private capital base or institutional structures to support it, much less the kind of expertise that entrepreneurial academic researchers would need to launch new products. In our observations, there may be a concentration of wealth in some countries, but investors may be reluctant to invest in a country if it has insufficient capital and talent infrastructure to support new technology development.

Funding mechanisms that comprehend the necessity of interdisciplinary and applied work will be crucial if translational approaches continue to dominate. The National Academy of Sciences recently acknowledged the need for more integrative science, stating that the new biology is an integration of many subdisciplines as well as the integration of physicists, chemists, computer scientists, engineers, and mathematicians in order to solve complex scientific and social problems (National Research Council 2009). For translational regenerative medicine to succeed, a more systems-based approach will be needed, that is, the ability to identify components of complex systems and understand how they work together. This requires knowledge from diverse fields not normally associated with stem cell research including materials science and biophysics, improvements on scaling up laboratory culture systems for biomanufacturing, better quality control mechanisms that may not be based on conventional biology, computational techniques for modeling cell behavior and tissue construction, and other integrative ways of approaching problems. Engineering, chemistry, and physics have become a critical part of tissue engineering and stem cell research, not only due to their contributions to "enabling technologies," such as culture systems, high-throughput techniques for screening molecules, biomaterial scaffolds and matrices, and process automation, but also in identifying physiologic and mechano-physical properties crucial to in vitro cell processes. Consequently, these fields are making major contributions to discovery science.

Companies producing research tools, culture systems, and reagents have been successful in the market. Ironically, they are not as successful in raising investment due to a lower ROI than therapeutic products (Rao 2009).

2.2.2 *Policy Variability*

The considerable variability in policies around the world has affected the procurement of biological materials (including cells, tissues, and genes), derivation of cell lines, data and materials sharing, and ultimate use of cell lines for both basic and clinical research (Isasi 2009; see also Appendix B). This was particularly true in the early years, when the variance in informed consent standards, derivation practices, or ultimate use restrictions for cell lines derived from embryos hindered sharing of materials and collaborations across national or even institutional lines (Caulfield et al. 2009; Elster et al. 2008; McCoy 2009).

The reason often named for the variance in policy stances on stem cell research is religion—more specifically, Catholic and evangelical protestant sects—which tend to object to embryo use in research and almost any science that "tampers with nature." It is important to note that the political influence of religion is diverse and the existence of increasingly politically assertive religion(s) in some parts of the world is a much broader phenomenon than what has been experienced with stem cell research alone (Toft et al. 2011). The role of religion in influencing public policy and law in various countries has not always been consistent over time. The relative role of religion in governance is in constant interaction with other economic, social, and political forces and activities are often connected to other social movements. Also, while some generalizations can be made, it is a mistake to presume that "religion" means opposition to science, as many commenters have. No group is so homogenous that there are not differing stances within each religion.

While the field has evolved, and the focus is shifting to nonembryonic sources, policy (and analyses of policy impacts) has continued to focus on constraints related to embryo-derived cell lines.[2] A number of emerging issues facing the field as a whole have yet to be addressed or are addressed unevenly in different regions, including data sharing, intellectual property issues, and potential ethical concerns about synthetic biology and gene-editing techniques, the use of epigenomic data, and more.[3]

[2] de Vries et al. (2008) track the proportion of articles devoted to HESC, cloning, and other topics, showing the dominance of embryo-oriented concerns of bioethics articles.

[3] Chapters 4 and 5 address the tensions between enforced data sharing (especially for genomic data) and capturing intellectual property. Given the debates on synthetic biology in Europe and, to a more limited extent, the USA, and the history of debates and policy implementation on gene therapy, it is remarkable that such issues have not been raised in stem cell research.

Efforts have been made to harmonize standards and regulations across political, ethical, and national borders. International organizations such as the International Society for Stem Cell Research (ISSCR) and the International Stem Cell Forum actively promulgate standards for protocols, guidelines for research ethics, and norms for human trials (ISSCR 2008). Still, with such differing historical and cultural backgrounds in various locales, this is nearly impossible to accomplish.

For all the discussion in science policy literature about harmonizing policy to facilitate development of a field, governments can make their own decisions about science and technology matters, weighting morality, religious traditions, economics (or perhaps more specifically, markets), public opinion (or perhaps projected constituency votes), or other pragmatic decision-making factors. It is not enough simply to acknowledge that there is significant variance in policies or attitudes toward regenerative medicine; rather, in order to navigate the differences, it is helpful to understand why particular social values become embedded in decision-making processes (cf Gottweis, Salter and Waldby 2009).

For some countries, there are competing incentives on the road to global competitiveness. Conforming to higher regulatory and research ethics standards promulgated by a few countries may provide greater credibility in the global scene, but following less restrictive guidelines may enable faster entry into clinical trials, faster results, and draw more capital than others. For-profit firms with investor expectations of a fast and high return on investment may be tempted to find less challenging regulatory and ethical environments to explore stem cell products. This is not new: pharmaceuticals have regularly been tested in resource-poor countries with fewer regulatory limitations. The potential danger to the product sponsor is acceptability of the data to more stringent review authorities in the USA or EU. For the research subjects in those locales (frequently the vulnerable members of society who otherwise have little health care), the risks of participating in experimental trials are multiplied, especially if there is no ability to care for them post-trial (Petryna 2009).

2.2.3 Stem Cell Tourism

More controversial are clinics advertising treatments which have not been proven through more rigorous standards of conventional oversight and accessed through established clinical researchers but, instead, are offered directly to the public (Regenberg et al. 2009). This growing phenomenon has been a concern due to potential risk to patients but also risk to the field as a whole, should patients be harmed. It also raises an important question about the pressure to do translational stem cell research and the rush to get into the clinic without first gaining important knowledge from discovery science.

The "hype" around stem cell treatments is at least partly to blame for the increasing number of clinics around the world offering treatments for a wide variety of disorders. The clinics are not always transparent about their processes, expertise, or even the type of cells being used. Many of these clinics (but not all) are in resource-poor countries which may not have stringent oversight or regulation over research or therapeutic practices. Patients—usually from wealthy countries—spend tens of thousands of dollars for procedures to travel to these sites, hence the term "stem cell tourism."

Medical tourism is not new; patients desperate to find treatments for rare or terminal diseases have long sought treatments outside of the bounds of conventional allopathic, scientifically proven medicine (Turner 2007). At the same time, there is a rapidly growing industry of clinics offering routine high-tech medical care in countries where the costs are more affordable than others, which may confuse health consumers in terms of which clinics offer proven treatments (ibid). This may make it more difficult for patients to discern legitimate trials from offers of treatments advertised by clinics which have not undergone rigorous review.

Not all sites recruiting patients are outside of North America and Europe.[4] One US-based web site positions itself explicitly as a patient advocacy organization, helping people to find a stem cell treatment and physician for members but goes further by to claiming entitlements to treatments as a self-determination and patient rights issue. From their web site: "We believe in the human voice and the human spirit. We believe that if people take care of their bodies and become their own advocates for determining what treatments are right for them that we can start an uprising for the general public" (Med Rebels, found at http://medrebels.org/about-us/). This is confusing for patients who seek treatments, particularly for disorders with few treatment options, particularly in the neoliberal public policy climate of the past two decades in many countries promoting patient self-care and "empowerment".

The current atmosphere pits cautious, risk-minimizing approaches to stem cell treatments against aggressive approaches framed as progressive and patient-consumer oriented. Countries restricting clinical use until extensive tests for safety are sufficient to proceed may appear to be blocking access to needed treatments (much like any other novel, first-in-human therapy), while others, promoting their services directly to patients, promise easy access to treatments that home countries refuse to provide for political reasons (Petersen and Seear 2011).

Furthermore, legitimacy is established when governmental authorities explicitly allow experimental treatments to proceed before evidence exists that they are safe and efficacious. The Italian Parliament, for example, is considering a new law making it legal to conduct such treatments in Italian public hospitals, outside of EU and Italian regulatory laws (Bianco et al. 2013). In Texas, new guidelines allow stem cell procedures as long as they are done for research, receive approval from an institutional review board (which could be private and profit-making entities, not attached to medical institutions), and patients must sign an informed consent form (Cyranoski 2012b; Park 2012).

While patients are warned against participating in treatments that have not been validated, clinics often use a narrative framework of offering hope and often post

[4] See, for example, the 2013 report on practices at US-based Precision Stem Cell at http://www.alsworldwide.org/documents/PrecisionStemCellReviewMarch192013.pdf and a subsequent blog-post discussing the report at http://www.healthintheglobalvillage.com/2013/05/06/precision-stemcell-selling-stem-cells-treating-individuals-with-als-as-human-guinea-pigs/.

positive testimonials on the Internet (ISSCR 2008; Lau et al. 2008; Lindvall and Hyun 2009). In the Italian case, public campaigns in favor of allowing treatments claimed that compassionate therapy was being denied to dying children if they were denied. In the Texas case, the debate was shaped by the governor, Rick Perry, who had himself undertaken autologous stem cell treatments and strongly supported allowing the procedure to continue. Such narratives frame the participation in experimental treatments as a patient right and autonomy issue and add credence to the effectiveness of treatments. At the same time, they may take advantage of the hopes and fears of patients in a vulnerable state.

The operation of questionable clinics with treatments that have not been validated (or worse, are based on pseudoscience) is disturbing, as patients may indeed be harmed. Yet focusing on the spectacle of stem cell tourism obscures an equally disturbing situation; that is, the phenomenon makes the environment far more difficult for legitimate researchers to navigate. The already-difficult decision about what sort of trials to execute and where, and how much preclinical data is sufficient before testing experimental treatments in humans, becomes far more sensitive in light of the spotlight on such controversial treatments.

2.2.4 Collaborations and Expertise

A number of organizations have arisen to galvanize research, help researchers to coordinate efforts, and garner public support. These have all taken different forms, in some cases being public-private partnerships and in others formed as not-for-profit advocacy groups. Such groups can sometimes bridge the gap when there is weak national leadership or there are no clear national strategies or where there is a need to bring patient groups together with researchers. A few examples include the Alliance for Regenerative Medicine (ARM) in the USA (a nongovernmental consortium of industry, university, and patient organizations which lobbies for favorable policy and regulatory environments and sponsors scientific exchanges), the Genetics Policy Institute, and the Tissue Engineering and Regenerative Medicine International Society-North America, among others. Internationally, groups include the International Society for Stem Cell Research (ISSCR), the Stem Cell Network (SCN) in Canada (which provides funding for scientific and policy research as well as conducting public education activities), the UK SCN (a publicly funded organization for scientific exchange and to promote commercialization), Stem Cells Australia (a recently reformulated effort focused on interdisciplinary collaborative interactions), and the Scottish National SCN (publicly funded for a fixed term and functioned as an advocacy and scientific community organization). The Japan Society for Regenerative Medicine was created in 2001 to promote research but has not been seen to be a strong coordinator of efforts across universities or link universities with industry; individual researchers have instead created the collaborations that exist. Chinese scientists also do not appear to have as much of a social infrastructure for collaboration and exchange as other countries.

We also observed a need to establish an adequate knowledge base or a way to acquire expertise through collaborations (cf Johnson et al. 2010). Collaborations for translational work are more effective where there are existing ties to industry and strong links between researchers and clinicians. For example, there are natural relations of industry and academia in places like Switzerland, which has a strong industrial base in pharmaceuticals. Also important is the capacity to conduct clinical trials. Locales where hospitals have appropriate equipment and expertise and where there is a critical mass of research expertise near clinical sites have an advantage. Research centers with close ties to clinics can not only obtain materials more easily (donated embryos for cell line derivation, cells for reprogramming, or bone marrow-derived stem cells) but have facilities in which patients can be treated and monitored.

As for expertise, we observed strong engineering components in stem cell research in Switzerland (EPFL), Germany, and the Netherlands. Japan has a strong engineering base, but integration of engineering and manufacturing with stem cell science is still in early stages. By comparison, engineering research in the USA and Canada is far more integrative with biology and addressing more fundamental discovery questions. This may be due in part to more experience with interdisciplinary research and training. Some countries adhere to strictly disciplinary training and ways of thinking about the role of specific types of expertise such as engineering, computational fields, or biology. They may utilize expertise from various fields but have not as readily taken up the kind of integrative, interdisciplinary research needed in regenerative medicine. In countries that have more at stake for economic development, there may be a tendency to use engineering as a practical way of building industry, rather than as a discovery science. For example, Portugal is capitalizing on historical experience in bioprocessing to move into regenerative medicine. It may also be that work on technical tools and aspects of research (e.g., the development of cell sorting and tracking systems or computational modeling) may be easier for policy-makers and funders to justify in an emerging, controversial field than research dealing directly with the use of embryos or genetic manipulations.

With this overview, we have introduced general patterns, which can be used to identify needs in this increasingly global field. In the next sections, we provide more situational analysis that helps to explain how regenerative medicine has developed the way it has within specific locales.

2.3 North America

This section provides an overview of funding and governance in the USA and Canada, adjacent countries with very distinct health-care and regulatory systems. Although there are many similarities, the two countries have approached the coordination of funding and support in very different ways. There is a considerable degree of cross-border collaboration, facilitated by the lack of language barriers and relative ease of access to researchers.

2.3.1 Canada

Canada occupies a unique position in the rise in regenerative medicine. Canada is home to James Till and Ernest McCullough, who helped define the fundamental properties of stem cells and thus catalyzed the development of stem cell therapies. A physicist and a physician, they studied the effects of radiation on bone marrow and demonstrated in 1963 the cardinal properties of stem cells, the ability to divide and give rise to other cells with similar developmental potential, and the ability to differentiate into the many cell (in this case) of the hematopoietic system.[5] Because their work had direct impact on cancer biology and therapeutics, it is not surprising that cancer research and regenerative science research emerged as a unified strength in Canada; that is, existing attention to cancer draws young investigators to the field, and funders would support this new, related area of research, focusing, initially, on bone marrow-derived cells and the hematopoietic system and expanding to other stem cell types with time and increased activity.

Stem cell research now occurs in several sites, dominated by large medical-academic centers in Toronto, Vancouver, and Montreal. However, there are few researchers relative to other countries, and they are spread across a large geographic territory. A stem cell network (SCN) was created and funded by the Canadian government to aid collaborations, support education and research in the field, and coordinate efforts, especially toward commercialization. The network covers costs that are not generally eligible for funding under federal research programs. The SCN has funded more than $42 million in interdisciplinary projects across Canada, resulting in 962 publications, 399 patent applications, 60 issued patents, and 43 licenses as of 2012 (Nerem et al. 2013).

Major sources of science funding in Canada are the Canadian Institutes for Health Research (CIHR) and the Natural Sciences and Engineering Research Council (NSERC), which funded the SCN. Through its funding and governance policies, the Canadian government places emphasis on how science can contribute to the Canadian economy, particularly in the past few years. For example, grant applicants are typically encouraged to discuss translational aspects of the impact of their projects as a part of their scientific proposals. Funds for basic science have not maintained the same growth rate as earlier years, although new funding mechanisms have been created to support academic-industry collaborations, including the Centres of Excellence for Commercialization and Research (CECR), created in 2007.

In an early effort to align research and commercialization efforts in Canada, the SCN set up a company, Aggregate Therapeutics, aimed at collecting stem cell IP and expertise from labs across Canada under one umbrella. Aggregate Therapeutics was eventually folded into MaRS, a Toronto-based organization created to promote and commercialize Canadian science.

A recent effort to commercialize regenerative medicine related technologies is Canada's Centre for Commercialization of Regenerative Medicine.[6] The CCRM

[5] Their work was published in 1963 in Nature (Becker et al. 1963).

[6] Coauthor Peter Zandstra is currently Chief Scientific Officer of the CCRM.

(funded by the CECR) was created as a not-for-profit organization to commercialize stem cell technology platforms based on Canadian researchers' strengths. Developed in close partnership with MaRS Innovation and the Canadian SCN, key platforms include reprogramming and engineering, biomanufacturing, and materials research. The CCRM is essentially a consortium of hospitals, universities, and industry members. Innovators can have their ideas and IP evaluated in terms of potential for commercialization and business model, and receive feedback, while members and potential investors have a right of early access to emerging IP. This arrangement is designed to be a more collaborative and transparent way of working on precompetitive research and a way to support research that is beyond the stage eligible for traditional basic research grants, but not yet at a stage for licensing or company creation. The CCRM may provide seed funding or co-funding with other organizations.

Key legislation affecting stem cell research includes the Assisted Human Reproduction Act, which governs embryo research in Canada. The Act prohibits buying or selling of gametes and embryos as well as human cells or genes for use in creating a human being. The Tri-Council Policy Statement (TCPS2), Canada's federal research ethics guidelines, contains a similar prohibition. The Canadian Institutes of Health Research updated guidelines specific to pluripotent stem cell research in 2010, which were adopted by the major funding agencies.[7] The guidelines address research ethics issues specific to the use of stem cells but also issues around commercialization, including a provision that donors must be informed if products from their biospecimens may ultimately be used commercially.

Possibly unique to Canada, a significant amount of funding has been designated for stem cell policy analysis, ethics training for researchers, and outreach to the public. A series of white papers and clear explanations of ethical and social issues is posted on the SCN web site, and several meetings have been convened on various aspects of stem cell research governance and ethics (see http://www.stemcellnetwork.ca/index.php?page=ethics&hl=eng).

In terms of regenerative medicine education in Canada, most graduate activity is centered around medical school or biomedical/bioengineering-related research programs in the larger academic centers (such as the Institute for Biomaterials and Biomedical Engineering at the University of Toronto). The CIRH supports a national Training Program in Regenerative Medicine, with online national and international courses and laboratory exchange programs. At the undergraduate and high school level, one strategy has been to educate high school teachers and key classroom leaders through the StemCellTalks program, a national stem cell biology outreach initiative in partnership with Let's Talk Science and the SCN. From a more translational perspective, the NSERC of Canada recently funded a Collaborative Research and Training Experience (CREATE) Program in RM Manufacturing, Materials and Mimetics (M3). Despite these initiatives, a coordinated national strategy for fundamental and translational (both clinical and manufacturing) regenerative medicine training remains to be developed.

[7] A summary of the guidelines can be found at http://www.cihr-irsc.gc.ca/e/42071.html.

2.3.2 *United States*

Ever since James Thomson became the first to successfully cultured nonhuman primate and then human embryonic stem cells, the USA has been considered as the leader in the field. The USA has strong labs and sufficient private capital and institutional support to sustain research.[8] Productivity (as measured by publications) is high in the USA, with an estimate of 38 % of world publications in stem cell research. One quarter of these had author collaborations with researchers in other countries (Luo et al. 2011). Yet the USA has no dedicated national strategy for stem cell research, and research and funding policy has been in flux since the early years of stem cell research.

In the absence of centralized federal leadership, some individual states have made stem cell research an explicit priority and have allocated budgets for this purpose.[9] Of particular note, California voters passed a state initiative creating a $3 billion fund for stem cell research at California institutions for a 10-year period. The organization created to fund and oversee the research is the California Institute for Regenerative Medicine (CIRM). Because the primary justification for its existence was to translate research into products beneficial to the taxpayers of California who voted in the initiative, the current focus of projects is on translational research and commercialization (Longaker et al. 2007). CIRM has agreements with the UK, Canada, and Japan to collaborate on research and recently contracted with a private company, Cellular Dynamics, to produce a bank of iPS cell lines for disease research, drug discovery, and other research.

The USA is characterized by more active public interest groups advocating for or against stem cell research than many countries. Advocacy and lobbying groups on both sides have actively worked to change public policy, influence public opinion, and attract funding. Most, but not all, opposition groups are conservative religious groups (American Right to Life Committee, Focus on the Family, etc). Many of these are more broadly engaged in American politics, particularly opposing abortion, and were key constituents in the mid-1990s political shift to the right, setting the stage for the President Bush era rulings on stem cell research. Stem cell research entered this climate in 1998 with James Thomson's successful creation of an embryonic stem cell line. Their activities do not end with HESC research, however; ongoing attempts to influence legislation will likely affect many forms of regenerative medicine-related research (see Chap. 1). Groups promoting stem cell research include the Alliance for Regenerative Medicine (ARM), which works to promote legislation as well as regulatory and reimbursement plans to create favorable environments for stem cell research and product development. ARM also focuses on attracting VC and other private and public funds to the field. The Coalition for the

[8] The USA has been extensively discussed elsewhere; therefore, only key points for comparison will be covered here. See, for example, Johnson et al. (2011), Lysaght et al. (2008), Rao (2009), and Salter and Salter (2010).

[9] States enacting stem cell funding mechanisms are listed at http://stemcells.nih.gov/research/pages/stateResearch.aspx.

Advancement of Medical Research (CAMR), a consortium of patient advocacy groups, scientific societies, and university research centers, also lobbied for increased federal funding.[10] The Genetics Policy Institute is a pro-cures organization that hosts a unique forum which brings together patients and disease advocacy groups with scientists and industry representatives to discuss issues facing the field.

After considerable public debate and activities from these groups, the federal government made a decision not to outlaw HESC research but to disallow funding with federal (taxpayer) money. The 2001 Presidential Statement by President Bush disallowed federal funding for the derivation and use of embryonic stem cell lines except a small number which were already in use.[11] The lines approved by the NIH were included in a registry, and a national stem cell bank was created to house and distribute the approved lines.[12] With little federal funding, the overwhelming majority of embryonic stem cell research was done with private funds. This is significant, because there was little oversight or transparency about derivation or research practices in privately funded research. Another direct result was that individual states began instituting their own rules. Some states made bans of HESC or cloning research more explicit, while others, as mentioned above, created mechanisms to support it. CIRM also developed its own governing body to review guidelines as well as protocols, creating an analog to NIH review processes.

Executive Order 13505 ("Removing Barriers to Responsible Scientific Research," 2009) replaced the 2001 policy, allowing federal funding for HESC research and creating a new set of ethical guidelines for research and derivation of new lines.[13] However, for both existing and new lines, proof of provenance was required before lines could be included in the new registry. Provenance information included demonstration that informed consent forms for donated embryos had appropriate language informing donors in greater detail about the disposition of their embryos. Because a number of lines had been derived from embryos donated years before embryonic stem cells were successfully cultured and came from a variety of public and private clinics in several countries and because there has never been consistency in informed consent language across these sites, provenance was extremely difficult to track. As a result, many of the gold standard lines could not be used in federally funded research projects for more than a year after the new rules went into effect. Researchers had to stop work or obtain approval to switch to another line if the line

[10] CAMR recently merged into the ARM.

[11] The statement on August9, 2001 attempted to make a compromise, by allowing limited federal funding for certain lines approved by the NIH as meeting the following criteria: They must have been derived with donor consent and without financial incentives; they must have come from embryos created for reproductive purposes but not used. Lines from 14 countries were initially included; ultimately only 21 were available to researchers for use, as many were of poor quality or had restrictions on use.

[12] The National Stem Cell Bank was housed within WiCell at the University of Wisconsin. Federal funding for the bank ended in 2010, but banking services continued as the WISCBank, which now distributes both embryonic and induced pluripotent stem cell lines.

[13] Also rescinded was Executive Order 13435 which opened funding for nonembryonic, alternative sources of stem cells. http://www.gpo.gov/fdsys/pkg/FR-2007-06-22/pdf/07-3112.pdf.

they were using was not already approved under the new registry, even if it had been accepted in the previous registry and approved by the NIH. Ultimately, new lines were approved, and there are more than 200 HESCs are on the current NIH registry (http://grants.nih.gov/stem_cells/registry/current.htm). In the meantime, a number of competing private and public banks proliferated around the world.

The change in policy, while permitting federal funding, meant a lengthy time for approval of long-used lines and uncertainty over which would be approved for use. Then, just as the bottleneck was easing, additional law suits attempted to block funding once the new guidelines took effect.[14] All of this made for considerable uncertainty for researchers and investors regarding the stability of funding and research policy. The impact on ongoing and future research was palpable, affecting international collaborations as well as work within the USA. [15] A survey of US stem cell scientists found that almost half of those using embryonic stem cells indicated that the ongoing uncertainty of national policy had a substantial impact on their research plans, but a number of those using human nonembryonic pluripotent stem cells also reported significant impact (Levine 2011).

In the USA, funding for most basic biomedical scientific research comes primarily from the National Institutes of Health, with additional funding for development phases coming from private industry. The National Science Foundation funds nonclinical science and engineering.

About $1.45 billion of NIH funds was spent on stem cell-related research in 2012, with most of this going to nonhuman, nonembryonic research, which includes all materials and techniques research. Only about 10 % ($146 million) of the total was devoted to HESC research, a relatively small increase since the pre-Executive Order amount of $88 million in 2008 (about 8 %).[16] By comparison, CIRM has a $3 billion fund over 10 years, New Jersey invested $380 million investment in a state

---

[14] Sherley v Sibelius U.S. Court of Appeals 11-5241. The case challenged whether embryonic stem cell research would violate the Dickey-Wicker Amendment which prohibits federal funding for any research which harms or destroys embryos. The final ruling favored allowing funding to continue (see Chap. 1).

[15] Levine, for example, surveyed 370 US researchers about the effect of the uncertain policy environment regarding embryonic stem cells (2011). The survey was taken after the 2009 US policy change allowing federal funding for HESC research and after Sherley v Sibelius. Of those surveyed, 18 % said that the resulting uncertainty meant they would either delay plans to begin HESC research, and 16 % said it would impede ongoing research (this group included those who had not previously used HESCs but were considering using them). Others reported shifting their research focus from HESCs to induced pluripotent stem cells. The disruption in recruiting new employees, consideration of a relocation, and disruption of collaborations were also mentioned as specific impacts, but these responses constituted fewer than 10 % of respondents. However, only 4 % said they would avoid using HESCs, and 3 % said they would consider relocating.

[16] A chart showing the pattern of NIH funding for stem cell research can be found at http://stemcells.nih.gov/research/funding/pages/Funding.aspx. The most recent NIH funding figures, including estimates for 2013, are found at http://report.nih.gov/categorical_spending.aspx. Umbilical cord blood is not counted in these figures, and the categories consist of research using keywords as defined by data mining algorithms, so may not pristeenly reflect actual research projects related to stem cell research.

stem cell institute, and Connecticut committed to $100 million over 10 years. An NIH Center for Regenerative Medicine was recently established to coordinate intramural research across centers. As mentioned above, the NCATS was also created to help facilitate the commercialization of research products.

In general the USA has a strong entrepreneurial base. There are incentives for academic researchers to commercialize their work, and many academic scientists have close ties to industry (see Chap. 1). Still, there is a gap between discovery and commercialization which has not been adequately addressed. Venture capital supported research to a limited degree initially, but after the financial downturn in 2008, investors became more risk averse (Rao 2009). The instability of private capital and difficulty in obtaining capital to get past scale-up and clinical trial hurdles, coupled with the stagnation or even decrease in federal research, is a major concern for technological innovation in the USA (ibid).

While entrepreneurial science is strong in the USA, so is discovery research. Engineering, for example, is more discovery-oriented than many other countries, where engineering is seen more as an applied science. Another strength of the USA is its educational infrastructure. Outstanding graduate and postgraduate education programs dedicated to regenerative medicine have been developed, and other countries encourage students to obtain training in the USA. Beyond laboratory training, however, students are provided with skills to be prepared to work in academia, government, or industry positions. As such, training programs are consistent with the National Science Foundation's goal of investing public funds to develop a strong scientific workforce capable of both basic discovery and commercialization.

2.4 Europe

The Lisbon Agenda, adopted by the European Council in 2000, was intended to make the European Union "the most competitive and dynamic knowledge-based economy in the world capable of sustainable economic growth with more and better jobs and greater social cohesion" by 2010.[17] The strategy was based on concepts of innovation and the knowledge economy, in which medical science technology plays a central role.[18] By most accounts, it was a failure, due in part to a lack of coordination among member states and the lack of political will to prioritize such an initiative among other pressing EU issues. The EU has since struggled to coordinate innovation policies, and there are key features that have kept it from being competitive with

[17] The rationale and goals can be found at http://www.consilium.europa.eu/uedocs/cms_data/docs/pressdata/en/ec/00100-r1.en0.htm.

[18] The term "knowledge economy" has been applied to the restructuring of economies through specialized expertise. In contrast to mass production or labor which characterized earlier agricultural and industrial manufacturing economies, it is knowledge—particularly in terms of engineering, science, mathematics, and information sciences—that drives the economy, and it is more global and interconnected in scope.

the USA, including comparatively weak ties between academia and industry; historically little venture capital investment, especially in high-risk innovations; and fragmented intellectual property laws (Hogarth and Salter 2010). The Innovation Union Strategy of 2010 was an attempt to rescue the aims of the Lisbon Agenda by coordinating innovation efforts across member states. Still, for regenerative medicine, major impediments remain, including limited availability of funding and intellectual property issues, especially in light of the exclusions from patentability which have constrained the research using embryos (see Noonan, Chap. 4, this volume).

Institutional structures within and across member states have also hampered development of stem cell-based regenerative medicine. Health technology assessments for new technologies, including regenerative medicine, are performed at the state level, with varying outcomes and recommendations. For example, the National Health Service structure in the UK is not a friendly environment for expensive, unproven innovations, which poses a problem for implementing stem cell products and therapies. As a result, the EU suffers from what has been called "the European paradox": excellent science without the capability of capitalizing on innovations and turning them into social and economic benefits (Hogarth and Salter 2010).

Research initiatives across the EU have been coordinated through the Framework Programmes (FP), and this mechanism is generally considered to be a more important source of funding than internal country funds. Each FP has supported the creation of networks of excellence as well as targeted programmatic themes. The current Seventh Framework Programme (FP7) runs from 2007 to 2013. There are specific programs within each, all meant to foster European scientific excellence.[19] Frameworks 6 and 7 have both provided funding for regenerative medicine projects (European Commission 2009). By mid-2010, about €187 million was spent on RM-related research (Kessler 2010). FP7 additionally encourages greater interaction between academia and industry. To address limits to research funding at the EU level, which had historically emphasized applied or industrial research, the European Research Council was created within FP7 in 2007 to fund peer-reviewed, "frontier" research broadly across the life sciences, engineering, and physical sciences. It is supported by the European Commission, with contributions from both member states, and associated states and is chartered through 2013.

The FPs have been an important way to draw additional disciplines such as advanced mathematical modeling into stem cell research. Cross-EU initiatives include European Consortium for Stem Cell Research (http://www.eurostemcell.org)

---

[19] There are additional programs, such as the Competitiveness and Innovation Framework Programme (CIP), Education and Training programs, and regional programs for competitiveness. The Innovative Medicines Initiative (IMI), somewhat similar to the National Institutes of Health "critical path initiative" in the USA, is one of the joint programs within FP 7. It was intended to identify new tools and new areas for drug discovery and related health technologies (Goldman 2012). It has a considerable budget; however, in controversial areas such as embryonic stem cell research, there are issues related to coordinating efforts across countries. It has also been criticized because of the large investment required by universities relative to other funding mechanisms (Sinha 2011).

and several coalitions around particular research problem spaces. Still, collaborations appear to be more common among researchers within a country, or perhaps with one other country, than Europe-wide collaborations. Nevertheless, A Europe-wide online database of HESC lines generated within Europe contains information on the origin and provenance of lines, plus genetic information, pluripotency, and other marker expressions (Elster et al. 2008).

There has also been a strategic emphasis on developing science clusters. However, these have not always worked well. As one commission report put it: "Europe does not lack clusters, but persistent market fragmentation, weak industry-research linkages and insufficient cooperation within the EU mean that clusters in the EU do not always have the necessary critical mass and innovation capacity to sustainably face global competition and to be world-class" (Cooke 2001; see also European Commission 2011). Although one aim of cluster research initiatives has been to aid poorer member countries, there is not enough concentration of scientific expertise in some regions to support sustained efforts; instead, efforts to support science infrastructures are focused mostly in well-established sites with expertise in biotechnology or drug development. These are supported at both the EU and state levels.

Disputes about policy and ethics regarding stem cell research in the EU have taken place within historical and political contexts about protections of individuals which differ somewhat from other regions. The aftermath of medical experimentation under National Socialism, as well as challenges in many areas of postwar science and technology (in particular, genetic engineering of organisms and crops), affirmed the need to adhere to a precautionary principle as the basis for policy decision-making. The precautionary principle, simply put, is the concept that when deciding whether to go forward with a new technology, and where there is no scientific consensus on whether it is harmful, the burden of proof that it is not harmful lies with those wanting to pursue that area of science or technology. Intended primarily for use for the prevention of environmental harms, it has been applied to a broad range of research, including medical research on humans. The precautionary principle has driven policy in the EU perhaps more than anywhere else: authorities consistently resort to this argument when attempting to write policy in response to controversies over emerging technologies such as genetically modified foods and gene transfer technologies (Dratwa 2011; Marchant and Mossman 2004).

Regenerative medicine policy has been entangled in this history. In 1989, amidst genetic engineering debates, proposed legislation in the European Parliament called for prohibiting gene transfer in the human germ line. The discussions set the stage for battles over intervening at the beginning of life as well as definitions and legal status of the human embryo. The term "human embryo" had been included in Article 6(2)(c) of Directive 98/44/EC of the European Parliament (see Chap. 4, this volume). However, while Article 6 prohibits "uses of human embryos for industrial and commercial purposes," which could be interpreted to be contrary to morality, no clear definition of an embryo was made either here or in the EU Council of 6 July 1998 statement on the legal protection of biotechnological inventions.

The ambiguity of definitions of the embryo became crucial when attempts were made to patent products of embryonic stem cell lines. Ultimately the European

Court of Justice determined that such products cannot be patented due to the so-called morality clause of the European Patent Convention which states that European patents will not be granted for innovations, the exploitation of which are contrary to *ordre public* (public morality) (EPC Article 53 (a)).[20]

FP6 negotiations resulted in Article 3 of European Parliament's amendments to allow use of supernumerary embryos but not embryos created from gametes for purpose of research. Nevertheless, member countries have authority to have their own guidelines. There is a broad range of country-specific policies, from the UK's Human Fertilization and Embryology Authority (HFEA) regulations of 2001, which permit use of embryo regardless of source, to Ireland's constitution which limits research, specifying the right to life of the unborn. In Belgium, France, and Denmark, research on embryos is allowed only if embryos left over from IVF procedures are used, and if the research relates to fertility or the prevention or treatment of disease.[21] In the UK, the HFEA is the agency which oversees all fertility procedures and has licensing authority for all research on donated embryos and gametes, including stem cell research. The 1990 Act which created the HFEA was amended in 2008 and again in 2011 to allow for research uses of embryos and the admixture of embryos containing human and nonhuman materials. Several countries, including India, have used the HFEA guidelines as a model for their own regulations (Bharadwaj and Glasner 2009).

A few features of individual countries help to illustrate the point about differing political, social, and historical contexts. The UK has been viewed as one of the more permissive countries, allowing not only HESC research but also the use of cybrids. There are a number of prominent research centers, including Cambridge University, Sheffield, and Imperial College London. The UK initially invested £29 million in SC research from 2003 to 2007 (Hogarth and Salter 2010). Most of the funding comes from the Medical Research Council (MRC), which funds basic and translational work, and stated regenerative medicine as a national priority for the UK economy and health care (Office of Life Sciences, UK Department of Health 2011). In 2009–2010, funding was at the level of £39 million per year, with plans to increase spending to £130 million in 2011–2014, including £100 million for a new technology innovation center.

The UK Stem Cell Initiative summarized the state of the field in the UK and made recommendations for action in a 2005 document called the "Pattison Report" (found at http://www.york.ac.uk/res/sci/events/FinalConfPres/Connolly.pdf see also MRC

[20] Greenpeace v Brüstle (see Chap. 4 and Gibney 2013)

[21] Belgium and France further specify that there must be no alternative therapy available (Loi relative à la recherché sur les embryons in vitro (Belgium); Loi no 2011-814 relative à la bioéthique (France)). Unlike most countries which allow embryo research up to 14 days of development, France requires that embryos be destroyed at 7 days. France updated its bioethics laws in 2011 to state that embryo research (including HESC) is only permitted in exceptional cases and is subject to approval by the Biomedicine Agency (see http://www.loc.gov/lawweb/servlet/lloc_news?disp3_1205402748_text). Articles 40–44 of the new law reiterate that research should only be done if there is likely to be a major medical breakthrough and there is no better alternative. Danish law derives from law on artificial fertilization, Lov nr 535 omkunstigbefrugtningsomaendretved.

2012). The lack of venture capital and other sources of funding for translational SC research was noted as a weakness, as well as the lack of clarity regarding IP and regulatory issues, and a consistent pattern of losing innovations to the USA in commercialization phases. Recommendations included the institutionalization of public-private partnerships and an increase in national funding for clinical and translational research.

Private capital in Britain has been slow to materialize (BIS 2011).[22] The Cell Therapy Catapult, launched in 2013, is one part of a national strategic initiative to grow new industries in Britain, aimed at bringing together academic and industrial partners.[23] It initially receives funding from the UK, with some support from the EU, but is intended to be sustained through public-private R&D collaborations.

Scotland is home to the Roslin Institute, an animal science and quantitative genetics research institute where Professor Ian Wilmut was first to succeed in using somatic cell nuclear transfer to create a sheep clone (Dolly). Stem cell research continues primarily in Edinburgh and Glasgow. A new Scottish Centre for Regenerative Medicine was created and funded by University of Edinburgh, Scottish Enterprise, the MRC, and the British Heart Foundation.

The organization of research in France is interesting because of its powerful patient advocacy groups and their direct involvement in influencing the direction of research. The Institute for Stem cell Therapy and Exploration of Monogenic diseases (I-Stem) was created in 2005 as a public-private collaboration between the French muscular dystrophy patient organization (AFM, which provided significant private, philanthropic funding), INSERM (French National Institute for Health and Medical Research), and the University of Evry-Val-d'Essonne (http://www.istem.eu/en/).[24] I-Stem, perhaps more than other collaborations, is tied to research on specific diseases, with a focus on rare genetic diseases, in particular, neuromuscular disorders. I-Stem has biobank stocks of patient cells, from which to make iPS lines, and works with Genethon, a clinical trial network for gene therapy also tied to the AFM.

Germany presents an interesting context in that there is a strong history of having leading basic and applied research institutes in Europe (resulting in a strong science and economic base). Germany took an early lead in clinical trials of stem cell therapies, aggressively pursuing cardiovascular therapies. Yet the history of human experimentation (resulting in strong protections of human dignity in the postwar constitution) created a difficult environment in which to pursue the use of embryos in regenerative medicine. As a result, Germany has the most restrictive policies in Europe (Stafford 2009). Interestingly, Germany allowed the import of HESCs while

---

[22] A report produced by the Department of Business Innovation & Skills of the Office of Life Sciences, Dept Public Health can be found at http://www.bis.gov.uk/assets/biscore/innovation/docs/t/11-1056-taking-stock-ofregenerative-medicine For a picture of patenting in the UK, see https://www.gov.uk/government/uploads/system/uploads/attachment_data/file/32456/11-1087-regenerative-medicine-patent-landscape.pdf.

[23] The catapult has produced a UK cell therapy clinical trial database, which can be found at https://catapult.innovateuk.org/documents/10726/1553967/CTC+UK+Clinical+Trials+Database/0451f336-4e2a-4907-a909-355e940b67b4.

[24] See Callon and Rabeharisoa for a detailed study of the way the AFM became a powerful stakeholder in national funding initiatives (2008).

banning the destruction of embryos on German soil.[25] Nevertheless, Germany's cross-departmental innovation strategy, the High-Tech Strategy (2006–2009) and High-Tech Strategy 2020 (2010–2013), include provisions for the support of non-embryonic stem cell research (German Federal Ministry 2007).

Core centers of translational regenerative medicine in Germany include the Berlin-Brandenburg Center for Regenerative Medicine, which has perhaps one of the most sophisticated structures in terms of integrating research projects with business organization. In addition to a matrix of research groups supporting diverse basic and translational projects, the center includes formal functions in business development and regulatory affairs, to help launch and sustain products successfully. Other research centers with strengths include the Fraunhofer Institute for Biomedical Engineering, which has historically focused on devices and technology, but increasingly works on regenerative medicine technologies, such as cell assays, tissue engineering (especially skin, liver, vasculature), biomaterials, and "lab on a chip" technologies.[26] Since its inception in 1987, it has focused on translational work and enhancing relations with industry.

Funding is jointly provided by the Federal and Länder (state) governments. The Deutsches Forschungsgemeinschaft (German Research Foundation (DFG)) is the major science funder and provides both project funding, capacity-building funds (for institutions or for centers of excellence), and Europe-wide collaboration funds. The DFG provided an estimated €17.9 million in 1999–2007 for stem cell research (individual grants) and €13.2 million for embryonic and tissue-specific stem cells under Priority Research Programmes plus €1.9m in 2001–2007 for clinical research.[27]

There is a strong engineering component to regenerative medicine efforts in the École Polytechnique Fédérale de Lausanne (EPFL) and Eidgenössisches Techniche Hochschule (ETH) Basel, in Switzerland. Switzerland is also in the enviable position of having a strong pharmaceutical industry and close ties between academic and industrial researchers with better possibilities for VC than other countries.

Denmark and Sweden have long-established histories of using human fetal tissue in research, particularly for Parkinson's disease (Kingman et al. 1992). This likely makes conditions easier in terms of regulatory and public support to conduct embryonic stem cell research. The fact that these countries are predominantly Protestant rather than Catholic may also play a role in how fetal and embryonic tissues are viewed for governance purposes. Sweden also has existing infrastructural and institutional capacities for translational research, which contributes to capacity building.

[25] Germany is a good illustration of the way that institutional histories shape governance of new areas of science. For a brief but fascinating analysis of the history of ethics decision-making bodies, see Jasanoff (2005, p.196).

[26] The Fraunhofer Institute for Immunology and Cell Therapy is also engaged in SC research. Information can be found at http://www.fraunhofer.de/en/research-topics/health-environment-nutrition/regenerative-medicine.html.

[27] Regenerationtechnologien für Medizin und Biologie—Beiträge für ein strategisches Förderkonzept (2007) found at http://www.biotechnologie.de/BIO/Redaktion/PDF/de/Studien/capgemini-regmed-2007,property=pdf,bereich=bio,sprache=de,rwb=true.pdf.

The Netherlands has been active in tissue engineering for many years and is the home to Eurotransplant, the central registry for transplant medicine. As such, there is considerable expertise in policy and logistics of tissue donation and exchange. The Institute for Regenerative Medicine supplies funding and support for tissue engineering and stem cell research. Ironically, there appears to be less integration with medical communities than might be expected and more in vitro work than translational cell therapy.

In terms of centers of education in Europe, a doctoral training program at Loughborough University stands out. In collaboration with the Universities of Keele and Nottingham, the program emphasizes skills in biomanufacturing but also provides training in several novel research platforms. In some countries, such as Germany, education has traditionally been strongly disciplinary-based, with little support or reward for interdisciplinary training or research. Strategies to deal with the inherently interdisciplinary work of regenerative science could be used to couple expertise from various labs or provide internships and training for junior researchers in relevant areas different from their home discipline.

2.5 Asia

While each country has its own political and cultural history, national health priorities, and ways of governing science, there are some regional commonalities in Asia when it comes to investments in biotechnology. Asian countries are attempting to assert themselves on the global scene as a part of nationalist projects to regenerate political identities in the wake of the turbulent twentieth century. The manufacturing industries of postwar developing Asia have given way to knowledge economies, in which biomedical research plays a significant role. Major national investments in science and technology are being made in China, Singapore, Taiwan, Korea, and other countries, and there has been renewed interest in alliances with non-Asian countries. The increased focus on science has turned scientists into highly visible national symbolic heroes, for example, Shinya Yamanaka in Japan and Hwang Woo-Suk in Korea. Before his fall from grace due to misconduct, Hwang Woo-Suk inspired a national postage stamp and a national campaign of volunteers to donate biological materials for his research.[28]

2.5.1 China

The rise of China on the global economic and scientific scene has been impressive. While NIH budgets have been relatively flat for several years, it has been estimated

[28] Hwang Woo-suk was a national celebrity and symbol of Korean resurgence in modern science after the announcement that he had successfully cloned human embryos. Later, it was discovered that he had engaged in scientific fraud and he was removed from his position (cf Gottweis and Kim 2010).

that Chinese investment in science and technology has increased 20 % per year. Venture capital (VC) investment grew rapidly after regulatory reforms in the 1990s began to open possibilities for private capital, and the "Patent law of the People's Republic of China" was passed in 2001. Foreign-based VC increased to 60 % of total VC by 2008 (Salter 2009b). Still, researchers have told coauthors of this chapter that they have little faith in the enforcement of Chinese intellectual property law, and this is seen as a significant barrier to commercialization.

Researchers funded by the Chinese Academy of Science (CAS) published 3.5 times as many papers in journals listed by the Science Citation Index (SCI) in 2009 as in 1998, and the number of papers published in the top 1 % of SCI journals (as judged by impact factor) was 12 times that in 1998 (Qiu 2011). The CAS has nonetheless urged researchers to shift their efforts away from quantity to quality; that is, rather than simply attempting to produce as many papers as possible, they should work toward genuine originality and innovation (http://www.cas.cn). Educating future scientists and retaining Chinese expertise is also a priority. Lab experience and education in the USA, Europe, and Japan are still strongly encouraged, which may be a problem for retaining well-qualified scientists in China. Programs such as the Thousand Young Talents Program have been created to recruit Chinese scientists to return to China and to recruit foreign postdoctoral students (http://www.cas.cn). Nevertheless, preventing brain drain is a significant problem and barrier to developing a top-notch research environment.

To understand how regenerative medicine research is unfolding in China, one must first understand the transformation of health care and scientific research within the post-Maoist economy. There has been a shift from preventive, collectivist health-care characteristic of the Mao era to more high-tech, innovative science that can compete on the global stage. However, China lacks the infrastructure needed for commercializing research or directing it toward national needs. Still in transition from a centralized economy, Chinese researchers and governmental authorities have less history of working with industry than other countries. There were simply no incentives to create new industries until program 863 (State High-Tech Development Plan), which began a transition to more capitalist forms of enterprise and incentivized scientists to get training in advanced research centers and then return to China to establish firms there. The 863 Program applies funding to critical areas of research in order to limit China's reliance on foreign powers (Huang et al. 2004; Thornley et al. 2011).

Perhaps because of a new desire to "catch up" on the global scene, China has invested more in translational research than basic. Yet "catching up" does not mean adopting Western or Northern globalizing ways of commercializing technologies. President Hu Jintao made this clear in a speech in 2006, when he called for China to create "a new path of innovation *with Chinese characteristics* and strive to build an innovation-oriented country" (quoted in Salter 2009a, p.402, emphasis added). Whatever is intended by "Chinese characteristics," China's Eleventh Five-Year Plan (2006–2010) emphasized the need to develop its own science and technology platform and capacity and exploit its own intellectual capital rather than relying on cheap labor to develop inventions of others (Ibid).

Salter points out that in the USA and other countries, venture capital is usually a source of management advice, recruitment, and fostering of talent. There is no such history in China; rather, individuals in companies relate to each other through *quanxi*, a system of mutual obligations and benefits between individuals and firms (Ibid). Trust or obligation established through such social relations affects investment patterns and responses to risk (high-risk products or firms). There are virtually no private foundations or sources of funds.

Also, there are tensions between China's desire to maintain tight government control, including financial protections, and the need to globalize. For example, venture capital normally looks for a 5-year exit plan by offering shares, preferably via an IPO. Markets are less fluid in the Shenzhen and Shanghai stock exchanges markets than this strategy allows, however, and government controls on the export of capital makes it hard to exit through other foreign exchanges (Ibid 2009).

Regenerative medicine is one of seven research priorities named by the CAS in its Innovation 2020 plan. Most funding for stem cell research comes from grants from the Ministry of Science and Technology in two programs, one for basic discovery science and the other for applied research. Approximately 100 million yuan (roughly US$12 million) was allocated between 2000 and 2005. The National Science Foundation of China (NSFC) is another major source of funds. Matching funds have typically come from local governments, in particular, Beijing and Shanghai (Murray and Spar 2006). Although estimates vary widely, and data reported by the CAS are incomplete or outdated, funding for stem cell research from 2000 to 2005 was approximately 300 million yuan (about US$38 million) and estimated to increase to about 400 million yuan in 2011 (Yuan et al. 2012; Murray and Spar 2006; Salter 2008). A national stem cell bank (primarily for HESCs) was also planned by the MOST (Xu 2008).

Research institutes at Beijing, Shanghai, Guangzhou, and Kunming are the pillars of the 2020 plan, and three science parks will be built in Beijing, Shanghai, and Guangdong to do translational research in informatics, biomedicine, and renewable energy (Qiu 2011). Stem cell research sites include Peking University's Stem-Cell Research Center and the Institute of Zoology at the CAS (Beijing) Xinhua Hospital (Shanghai) and Xiangya Medical College (Changsha), the National Institute of Biological Sciences (Beijing), the Shanghai Institutes for Biological Sciences of the CAS, and the Guangzhou Institute of Biomedicine and Health of the CAS (Guangdong). Large pharmaceutical companies such as Pfizer and Johnson & Johnson, as well as smaller biotech companies, have actively engaged with researchers in these sites and have initiated clinical trials under the "Developmental and Reproductive Research Initiation," organized and sponsored by the Ministry of Science and Technology of China.

An example of the new arrangements is the Bieke Biotechnology Company, a company using cord blood and bone marrow-derived stem cells to treat a variety of disorders. Beike Biotech was the first Chinese company to receive accreditation for cord blood and bone marrow-derived stem cells. Bieke has a network of clinics and researchers throughout China but is based in Shenzhen, a special economic zone (SEZ). SEZs were created under the reform policies of Deng Xiaoping to attract foreign investors (especially ex-patriot Chinese) and as such are exempt from many of the usual province regulations. Cells can be shipped to clinics within the network, and if one province does not allow a type of procedure, it can be easily moved to another site.[29] A web site, China Stem Cell News (found at http://www.stemcellschina.com), billed as a news and information service for patients wanting to know about stem cell research, steered potential patients to Bieke for treatments. The online tool became a primary recruitment mechanism for patients around the world with a variety of disorders to receive treatments in China.

The practice of direct-to-consumer advertising for patients, using treatments with little documentation, makes China is a key target for international criticism about clinics which offer stem cell treatments which have not been proven to be safe or efficacious. In fact, China is sometimes derogatorily referred to as the "wild east" of regenerative medicine, as though there are no rules or oversight. Chen argues that it is not that there are no rules and a passive populace; oversight does exist, but in a way that researchers are somewhat easily able to navigate or bypass regulations (Chen 2009; see also Rosemann et al. 2013).

The Beijing Ministry of Health (MOH) Medical Ethics Committee and Southern Chinese Human Genome Research Centre Ethical, Legal, and Social Issues Committee (ELSI) proposed ethical guidelines for HESC research in 2001, including the establishment of a new organization to centralize ethical management of stem cell research in China. In 2003, the Ministry of Science and Technology and MOH issued ethical guidelines for HESC research. However, while there are local ethics committees, there is no centralized management to date (McMahon et al. 2010; Zhang 2012b). As a result, there is little infrastructure for oversight or penalties for noncompliance with guidelines. The MOH did make a provision regarding cloning, allowing somatic cell nuclear transfer (cloning for therapeutic purposes) but not cloning for reproductive purposes. The guide is referred to as the "Four No's": Under no circumstances will human reproductive cloning experiments be endorsed, permitted, supported, or accepted.[30] The only source of materials for research officially permitted includes supernumerary blastocysts after in vitro fertilization (IVF) procedures, fetal cells from accidental spontaneous or voluntarily selected abortions, parthenogenetic split blastocyst obtained by somatic cell nuclear transfer technology, or voluntarily donated germ cells. Still, some have argued that the history of state-forced abortions and birth control makes it easier for the State to intervene in reproductive issues, including potentially making human oocytes available for research. In fact, one Chinese researcher interviewed claimed that oocyte donation was little different from blood donation and that oocytes were easily acquired from cooperating IVF clinics without special consent (Sleeboom-Faulkner

[29] The first patient, an American with amyotrophic lateral sclerosis, was offered free treatment in 2005. The procedure was performed at Nanshan Hospital in Shenzhen. Cells from a lab in Zhenzhou were used. Another patient wanted to have cells injected directly into the lamina for multiple sclerosis, but physicians at Nanshen refused; the patient was sent to another province where a clinic was willing to perform the procedure (Song 2011, p 147).

[30] People's Republic of China Ministry of Health (PRC MOH) guidelines for clinical use of biomedical technologies is available at http://www.mmoh.gov.cn/publicfiles/business/htmlfiles/mohyszs/s3585/200903/39511.htm.

2013). There is also a large pool of potential research subjects for all forms of medical research, with relatively easy access compared to other countries.

Responding to the international concern about unproven therapies being offered, in May 2009, the Chinese MOH classified stem cell treatments as high-risk medical technologies, requiring the approval of an audit board. Sponsors using stem cells were asked to register their research and clinical activities, the source of the stem cells, and ethical procedures. The Ministry also asked local health authorities to stop any unapproved clinical uses and called for a nationwide moratorium on new clinical trials for stem cell therapies on 10 January 2012 (Durfee and Huang 2012). In July 2012, 50 clinics were selected to conduct approved stem cell trials or treatments. At the time of this writing, other clinics continue to operate, leading many to believe the practice is not being well regulated.

On 1 May 2009, the MOH promulgated the "Management Measures for the Clinical Use of Medical Technologies," a regulation that classified a range of new medical technologies and procedures into three categories. Stem cell transplant technology was grouped under category III, which included technologies considered as risky, ethically controversial, and in need of clinical verification (Qiu 2011). To implement the regulation, the MOH assigned five institutions, among them the Chinese Medical Association, the Chinese Hospital Association, and the Chinese Doctors Association to take the lead. Clinics using stem cells were supposed to register with these institutions, and licenses would be granted on the basis of assessment criteria and approval by review and inspection committees (Chen 2009, p. 271). In practice, this regulation has not yet been implemented, particularly in district and military-owned hospitals, partly due to disagreements about how the policy should be implemented (Chen 2009). On 6 January 2012, the MOH issued a regulatory document called "Notification on Self-Evaluation and Self-Correction Work regarding the Development of Clinical Stem Cell Clinical Research and Applications." The four stages of this approach are self-evaluation (zicha), self-correction (zijiu), re-certification (chongxin renzheng), and standardized management (guifan guanli) (Rosemann 2013). Still, it is not evident that regulations are being followed (Cyranoski 2012a).

Priscilla Song situates medical tourism for stem cell research in China within the context of changes in political economy of health care and market reforms in China (Song 2011). Under Ding Xiaoping, funding for state-owned hospitals was significantly cut. In the transition toward what has been referred to as "socialism with Chinese characteristics," there was a mix of decentralization and central control of many economic sectors: Hospitals were allowed to raise fees on some services but were mandated to have price controls on services defined as "essential." Hospitals responded by moving toward high-tech, lucrative services and creating elite wards, which they reserved for wealthier clients (including foreign) or leased to companies for clinical trials. While there are national regulatory limits to companies (especially foreign firms) accessing patients this way, Song observed that companies circumvent legal constraints on such arrangements by working through universities and local governments which do not go by national rules. Companies can thus gain legitimacy and access to clinical facilities and patients (Song 2011, p.143).

It is important to note that legitimate trials are being conducted at the same time as questionable ones. This creates even more of a dilemma for China as it strives to achieve world status in regenerative medicine.

2.5.2 Japan

In 2003, The Japanese Ministry of Education, Culture, Sports, Science, and Technology (MEXT) wrote a white paper in which regenerative medicine was named as a priority strategy for science and technology.[31] As an economic stimulus program, the RIKEN Institute (under MEXT) led an initiative to support regenerative medicine in 2003–2008. After the Yamanaka discovery of iPS cells in 2007, MEXT added about ¥ 1 billion to the budget for RM for 2008–2012. The Japan Science and Technology agency (also under MEXT) has been a major funder of IPSC research. Additionally, the Ministry of Trade, Economy and Industry (METI) also provided funding for industrial applications of stem cells including cell culture systems and automation systems, cell sheet manufacturing, and measuring devices with a budget of about ¥5.5 from 2008 to 2014 (Japan Science and Technology).

Japan was the earliest Asian nation to industrialize, developing a strong base of manufacturing industries and expertise, including electronic and mechanical engineering. Significantly, Japan has strengths in robotics and optics, which will be important to developing tools for automated bioprocessing and scale-up as well as cell tracking and other imaging uses. This is not surprising, from the history of auto industry, photographic equipment and supplies, and electronics devices manufacturing. Some of the firms in these industries are retooling into biosciences as markets for conventional products shrink. For example, the Fujifilm Corporation, a company which previously made photographic film, is utilizing its expertise in chemistry to enter the tissue engineering field by partnering with Japan Tissue Engineering. They aim to build on knowledge of collagen and polymers, as well as mass production techniques, to make scaffolds and microspheres. Olympus (a maker of cameras and microscopes) is expanding to live cell imaging, and Nikon is now selling specialized stem cell equipment, including automated cell culture and monitoring stations. In addition to engineering expertise, there is considerable expertise in developmental biology and transgenics. There is a solid pharmaceutical industry presence, and a number of these firms are starting to work with stem cells. The pharmaceutical industry in Japan appears to be somewhat more risk averse, in entering new areas, in contrast to other Asian countries like China, which has engaged in high-risk strategies.

Japan has almost exclusively pursued IPSCs rather than HESCs. Some might argue that the successful research by Shinya Yamanaka on induced pluripotent cells has been a primary influence on direction for the nation. Others might argue that cultural preferences not to use embryos is the reason, although there are somewhat conflicting accounts of the extent to which this is true. The dominant religion,

[31] http://www.mext.go.jp/english/whitepaper/1302732.htm.

Buddhism, would likely be less concerned with the beginnings of life than the end. The use of stem cells from aborted fetuses has been allowed since 2004, and research on cell lines from embryos has been allowed since 2001.[32] Sleeboom-Faulkner's research in Japan revealed that there has been little public discussion about embryonic stem cell research but suggests that contrary to assumptions that the status of the embryo is unimportant in Japan, embryos do have significant moral status (2008). Indeed, the Bioethics Committee of the Council for Science and Technology Policy (CSTP) defined an embryo as *seimei no myooga* (the "germ" of life). Japan also has a morality clause regarding patenting of products deriving from the destruction of embryos and was the first country to utilize this concept.[33] Sleeboom-Faulkner argues that permission to use embryos and fetuses has less to do with religious views than with political priorities of building a large-scale infrastructure for science in the beginning of the twenty-first century, such as the Millennium Project (Sleeboom-Faulkner 2008).[34]

While therapeutic applications are being pursued, ShinyaYamanaka, the Nobel Prize-winning scientist and inventor of iPS cells, has called for a national priority to be in areas other than cell therapies: "My goal is for 20 % or so of iPS cell applications to be in regenerative medicine and the remaining 80 % to be in finding the causes of diseases and developing drugs. Japan is making advances in researching regenerative medicine, but we are lagging behind the West in other applications" (Yamanaka, quoted in Oiwa 2013; see also Cyranoski 2012c).

To build an infrastructure around iPS cell technologies, Japan is creating a national biobank for iPS cells. The Health Ministry in 2012 approved the addition of umbilical cord blood samples from the eight national cord blood banks to other cell stocks that might be used to derive cell lines. Notably, fewer subjects would be needed to make a biobank in Japan, because a smaller number would still represent a majority of the population due to the relative genetic homogeneity and similarity of HLA profiles.[35] Dr. Yamanaka plans to create 75 iPS cell lines that could be

---

[32] The derivation of new HESC lines requires a two-stage approval process (Institutional Review Board and Ministry level review). There is also considerable structure regarding research ethics, including a requirement that each institution have bioethics and technical training courses approved by the Ministry.

[33] Chapter 2, Sect. 32, concerning "unpatentable inventions" contains the clause: "the inventions liable to contravene public order, morality or public health shall not be patented." See http://www.wipo.int/clea/docs_new/pdf/en/jp/jp006en.pdf. The language of the law allows some room for interpretation about respecting an embryo while permitting it to be used to create life-saving therapies.

[34] The Millennium Project, a national initiative created in 1999 as an economic stimulus project, was intended to develop an infrastructure for science and technology. A public-private collaboration, it was jointly sponsored by the Ministry of Education, Sports, Science and Technology (MEXT); the Science and Technology Agency (STA); the Ministry of Health Labor and Welfare (MHLW); and the Ministry of Economy Trade and Industry (METI), linking education, commerce, and R&D into a networked initiative.

[35] Three key genes that code for cell surface proteins involved in the immune response (human leukocyte antigens, or HLA) must be matched to prevent possible rejection of the cells in the recipient. Banked cord blood in Japan will have already been characterized for HLA.

matched by 80 % of the population. One possibility for the Japanese citizenry is the stockpiling of cell lines, particularly blood cells, suitable for the majority of Japanese for emergencies (interview with Yuri Oiwa 2013; see also Cyranoski 2012b). With the history of radiation poisoning and subsequent bone marrow depletion in affected citizens in World War II, and again with the Fukushima nuclear disaster, coupled with a rise in blood-related cancer rates, this is not a surprising strategy. Recently acquired government-sanctioned access to the cord blood cells will facilitate the work but raises questions about donor consent, since the cells would be used for purposes other than what was specified to them at the time of donation (see also Chaps. 6 and 7 regarding problems of informed consent in cell-based research).

The MEXT is responsible for the enforcement of guidelines regarding stem cell research. Guidelines in Japan are guided by an Expert Panel on Bioethics, reporting to the CSTP, which has a reputation for being cautious and taking time to make decisions. Recently, the Panel did recommend to relax guidelines on admixture of human cells into animals (Normile 2013). Previously, this was allowed in vitro but not in vivo. The change was announced just prior to an announcement of the successful growth of human liver tissue from iPS cells, which began in vitro but needed to be completed in vivo to more properly form three-dimensional structures. Guidelines for Derivation and Utilization of Human Embryonic Stem Cells were created in 2001 and amended to relax requirements in 2009. Clinical trials must go through the Ministry of Health, Labor and Welfare (MHLW) for approval. Guidelines on clinical research using human stem cells were created in 2006 and amended in 2010.[36] Paragraph 5 of the guidelines specifies medical conditions in which stem cells (iPS or ES) may be used, which are restricted to life-threatening illnesses, treatments for which stem cells are expected to produce significant improvements over existing therapies, and the benefits are expected to outweigh the risks. Interestingly, the guidelines also specify the kinds of expertise that should constitute an ethics review committee, including expert(s) in molecular biology, cell biology, genetics, clinical pharmacology or pathology, law, ethics, and clinicians in the area for which a therapeutic protocol is being developed, but not stem cell researchers (paragraph 8).

In terms of education and training, only four Asian universities are ranked among the top 100 in the world, and all of them are in Japan.[37] Fewer Japanese students are studying abroad, in contrast to China and India.[38] Key sites for regenerative medicine research include the University of Tokyo and the Institute of Advanced Biomedical Engineering of the Tokyo Women's Medical University. The institute also has links with other Japanese universities.

---

[36] These can be found at http://www.mhlw.go.jp/english/policy/health-medical/medical-care/dl/guidelines.pdf.

[37] These are the University of Tokyo, Kyoto University, Osaka University, and Nagoya University.

[38] The number of Japanese college students studying overseas dropped by 28 %, from 82,900 in 2004 to 59,900 in 2009, according to figures from the Ministry of Education, Culture, Sports, Science and Technology. In the USA, the number of Japanese scholars has fallen to half of what it was during the peak year of 1997 (Asahi Shimbun; see http://ajw.asahi.com/article/behind_news/people/AJ201210100003).

2.6 Discussion

As we have shown, regenerative medicine has developed differently in global regions, but not only because of funding or differences in policies. Rather, regenerative medicine has emerged within historical and political moments particular to regions and countries.

Research is flourishing or being constrained in part due to policies, but this entails policy regarding intellectual property and investments in science more broadly, not just policy regarding stem cell research. Research ethics policy regarding clinical trials or the use of embryos significantly affects what research may or may not be done, but so does policy about foreign investments or data and materials sharing across labs or national borders. Stem cell-specific guidelines in some areas are made to be similar to other countries for interoperability, or to assert international credibility, whereas in other areas, guidelines adhere to country-specific, long-held cultural ideas.

There are other influences: in some countries, advocacy groups and patient organizations searching for cures for particular diseases have pushed governments for experimental stem cell treatments, while in others, patients take matter into their own hands and depart for countries offering treatments, even without substantiated evidence of effectiveness and safety. The health-care systems into which stem cell therapies and other products will go also makes a fundamental difference in whether or not technologies, once translated, are taken up by payers and incorporated into clinical care under budgetary strain or in transition from a central economy.

Funding amounts do matter, but the source and type of funds, and the conditions under which they are provided shape what sort of research goes forth (or not). Where funds are tied to specific economic goals, they are likely to support commercialization efforts or the development of enabling technologies more than discovery research. There also may be short-term thinking, due to political exigencies or the conviction that choosing to invest in the development of a particular technology over another may provide short-term economic benefits or might be more politically digestible. Where one approach to research is controversial, public money will be shunted to approaches less likely to cause debate. When funding comes from venture capital or existing industry sources, products with a better return on investment will develop, but those for smaller markets or rare diseases will have to find other sources of support. Also, funds from industry or venture capital will likely be less likely to flow to projects requiring longer development times.

One thing is clear: research follows the money, and the money appears to be in all locales toward translational research and more commercially viable products and therapies than basic discovery science. However, as one coauthor (David Schaffer) put it, translational research cannot exist without something to translate. That is, commercialization only makes sense when we have high-quality science to commercialize, as Peter Zandstra adds. Funding cuts in basic discovery research may thus result in an empty translational pipeline in the future. The question then becomes, can recommendations be made for future funding and research policy to "lift all boats" of regenerative medicine research? If so, how can the field as a whole be served while preserving national or regional priorities for science, science education, the economics of science, and ultimately, health care? Which funding mechanisms would be most effective to draw together interdisciplinary perspectives needed for translational RM, and which would encourage collaborations across laboratory and national borders in order to leverage and capitalize on existing cores of expertise?

There are several possible ways of reorienting research policy, including a clinical strategy, a research organization strategy, a translational strategy, or a hybrid of these and other strategies:

- Are there clinically strategic ways of prioritizing types of research based on a disease-oriented basis (e.g., matching disease incidence with expertise in that area)? Currently, health needs in countries are not necessarily well matched with types of research being pursued. Alternatively, might it be more productive to support translational research where natural relations with clinics already exist, or should policies be directed toward bridging the gap between basic researchers and clinical practitioners? Should countries prioritize certain health needs in their federal funding initiatives or should this be left to private sources in locales where a strong patient advocacy base exists around a particular health need?
- Are there natural venues to collaborate around specific research problems or around a technique or approach? (Examples of emerging areas could be computational methods, or bioprocessing/biomanufacturing, reprogramming, or gene-editing techniques). If so, how can communities of interdisciplinary expertise be supported? How can funding agencies be made to understand and support such efforts (even internationally) in addition to conventional funding mechanisms?
- Where are opportunities to develop better working relations between academic researchers and companies that could develop nascent innovations, and what form of public-private partnerships might be most effective?

The WTEC report, aimed at American competitiveness, concluded that funding agencies should establish interagency programs for interdisciplinary stem cell research, joining engineering, biology, and computational scientists, among others. It also recommended mechanisms to support academic-industry partnerships, particularly those with innovative translational models. Additionally, grant programs which allow for collaborations across countries would leverage strengths across labs and national research programs in much the same way as intranational networks have done (see also US DHHS 2006).

Whatever approach is employed, there is a need to address the current disconnect between expertise, resources, and political will in many locales. A good deal of expertise exists in pockets around the world, and niches have developed in which experts interact within and across disciplines that may regulate the fate of stem cell research. Yet there is a need to identify and then build up expertise needed for the long term. That includes not only knowledge needed for translation and scale-up but also regulatory and business infrastructures. Microenvironments may work well to advance particular areas of research, but better matching of resources, expertise, cGMP facilities, and access to clinics will be needed for translation.

We have described a few experimental models created to accomplish translational stem cell research, including the CCRM in Canada, the Berlin-Brandenburg Centre in Germany, and the recently introduced Cell Catapult in the UK. Such novel models involve unique public-private partnerships and will bring together academic and industrial researchers. Such models are still experiments in process, and it remains to be seen what may be the best form. What is also needed is incorporation of more clinical practitioners to provide a better understanding of disease mechanisms and clinical picture of patients' actual needs, in addition to providing strategic access to patients for potential trials (Johnson et al. 2007). The mundane, less-studied, less-funded, yet crucial piece of clinical translation involves design of cell preservation and delivery methods, biomanufacture and scale-up, and other processes needed to make a therapeutic procedure workable.

In the future, translational centers will also need to incorporate the burgeoning information coming from genomics and data analytics to develop in vitro models of disease and conduct population level analyses in addition to personalized approaches. Countries with established information technology industries and the infrastructural ability to connect health, population, and bioscience databases may be able to capitalize on these strengths.

At the same time, a number of groups are reformulating training and education to focus on the interdisciplinary field of regenerative medicine rather than maintaining silos of disciplinary expertise. Loughborough University, for example, in collaboration with the Universities of Keele and Nottingham, created a program emphasizing skills in biomanufacturing in addition to training in several novel research platforms. Interdepartmental doctoral programs are arising to join disciplines, or in some cases, link basic and clinical sciences.

Individual institutions are beginning to innovate new ways to deliver education, including programs targeting international audiences. Stanford University's biomedical engineering design courses in India, formed as a collaboration between Stanford, the Indian Institute of Technology, and the All India Institute of Medical Sciences (http://biodesign.stanford.edu/bdn/india/), is one example among many others where universities partner bilaterally with institutions in China, India, Africa, and other resource-poor countries. Other universities are creating focused, international short-course training programs (such as Georgia Tech's new course in biomanufacturing). On a broader scale, new forms of open online courses are being implemented by universities as a part of a movement to make higher education more accessible to larger numbers of people rather than traditional campus-based, direct-interaction models. An extension of distance education, the concept allows learners in any locale (including resource-poor countries) to enroll in top-level courses and obtain college credit.[39] Although these forms are somewhat controversial, and there are many problems to work out, some version of such versions in stem cells and regenerative medicine may accelerate transmission of knowledge in these rapidly expanding fields from a small number of centers of excellence to the broader international community.

Future empirical research will provide more fine-grained analysis than we have been able to present here. In particular, it would be helpful to analyze disease incidence and to what extent existing stem cell research programs match patterns of health needs within locales. Longitudinal analysis could provide insight into the extent to which funding, or the creation of research communities around a technique or research problem, or hubs of interdisciplinary collaborations attract investigators to the regenerative medicine space, and whether they disperse when venues for collaboration decline. For our purposes in this chapter, we provided an overview with which to consider future policy directions and raised questions for further exploration.

References

Becker, A. J., McCulloch, E. A., & Till, J. E. (1963). Cytological demonstration of the clonal nature of spleen colonies derived from transplanted mouse marrow cells. *Nature, 197*, 452–454.

Bharadwaj, A., & Glasner, P. (2009). *Local cells, global science: The rise of embryonic stem cells in India*. New York: Routledge.

Bianco, P., Barker, R., Brüstle, O., Catteneo, E., et al. (2013). Regulation of stem cell therapies under attack in Europe: For whom the bell tolls. *EMBO Journal*. doi:10.1038/emboj.2013.114.

BIS. (2011). Taking Stock of Regenerative Medicine in the United Kingdom, (July). UK Department of Business, Innovation and Skills (BIS) and UK Department of Health. Retrieved March 13, 2013, from https://www.gov.uk/government/uploads/system/uploads/attachment_data/file/32459/11-1056-taking-stock-of-regenerative-medicine.pdf.

Brindley, D. A., Reeve, B. C., Sahlman, W. A., Bonfiglio, W. A., Davie, N., Culme-Seymour, E., et al. (2011). The impact of market volatility on the cell therapy industry. *Cell Stem Cell, 9*, 397–401.

Bruton, G. D., Fried, V. H., & Manigart, S. (2005). Institutional influences on the worldwide expansion of venture capital. *Entrepreneurship Theory and Practice, 29*(6), 737–760.

[39] Examples include massive open online courses (MOOCs) such as Coursera, a for-profit company that initially partnered with Stanford University, the University of Michigan, University of Pennsylvania, and Princeton University. Others include Kahn University (a not-for-profit model supported by Google Project 10 and the Gates Foundation) and edX (a not-for-profit partnership of Harvard, MIT, and the École Polytechnique Fédérale de Lausanne in Switzerland). See also Leber (2013). The jury is out on the value and effectiveness of such unconventional offerings. Many universities are rushing to develop similar offerings, hoping to capture revenues during a time of declining public and private support of higher education, but there are numerous hurdles to launching and maintaining courses, including standardization and quality control of course content, licensing, and the time-intensiveness for high-quality, busy professors with little expertise in teaching international audiences to prepare online materials and monitor and evaluate students coming from radically different backgrounds and cultural settings. Language translation is also an issue; currently, most courses are in English only.

Burrill and Company. (2012). Retrieved May 2012, from http://www.burrillandco.com/content/news/PR-bcmonthly-1-3-12F%20v2.pdf.

Callon, M., & Rabeharisoa, V. (2008). The growing engagement of emergent concerned groups in political and economic life. Lessons from the French Association of Neuromuscular Disease Patients. *Science, Technology & Human Values, 33*(2), 230–261.

Caulfield, T., Zarzeczny, A., McCormick, J., Bubela, T., et al. (2009). The stem cell research environment: A patchwork of patchworks. *Stem Cell Reviews and Reports, 5*, 82–88.

Chen, H. (2009). Stem cell governance in China: From bench to bedside? *New Genetics and Society, 28*(3), 267–282.

Cooke, P. (2001). New economy innovation systems: Biotechnology in Europe and the USA. *Industrial Innovation, 8*(3), 267–289.

Durfee, D., & Huang, S. (2012). China stops unapproved stem cell treatments. *Reuters*. Retrieved November, 2012, from http://www.reuters.com/article/2012/01/10/us-china-health-stem-cell-idUSTRE8090GA20120110.

Cyranoski, D. (2012a). China's stem cell rules go unheeded. *Nature, 484*, 149–150.

Cyranoski, D. (2012b). Stem cells in Texas: Cowboy culture. *Nature, 494*(7436), 166–168.

Cyranoski, D. (2012c). Stem cell pioneer banks on future therapies. *Nature, 488*(7410), 139. doi:10.1038/488139a. Accessed November 2012.

de Vries, R., Oerlemans, A., Trommelmans, L., Dierickx, K., & Gordijn, B. (2008). Ethical aspects of tissue engineering: A review. *Tissue Engineering: Part B, 14*(4), 367–375.

Dratwa, J. (2011). Representing Europe with the precautionary principle. In S. Jasanoff (Ed.), *Reframing rights: Bioconstitutionalism in the genetic age* (pp. 263–286). Cambridge, MA: MIT Press.

Elster, A., Damaschun, A., Kurtz, A., Stacey, G., Arán, B., Veiga, A. et al. (2008). The changing landscape of European and international regulation on embryonic stem cell research. *Stem Cell Research, 2*, 101–107. doi:10.1016/j.scr.2008.10.003.

Etzkowitz, H., & Leydesdorff, L. (2000). The dynamics of innovation: From national systems and 'mode 2' to a triple helix of university-industry-government relations. *Research Policy, 29*(2), 109–123.

European Commission. (2009). European research projects involving stem cells in the 6th Framework Programme. Retrieved December 20, 2013, from http://ec.europa.eu/research/fp6/p1/stemcells/pdf/stemcell_eu_research_fp6_en.pdf#view=fit&pagemode=bookmarks.

European Commission. (2011). Report from the Commission to the European Parliament, the Council, the European Economic and Social Committee, and the Committee of the Regions. State of the Innovation Union 2011.

German Federal Ministry of Education and Research (BMBF). (2007). *Regenerative technologies for medicine and biology—Contributions for a strategic funding concept*. Berlin: Capgemini Deutschland.

Gibney, E. (2013). Court allows patenting of stem cell technologies. Retrieved January 10, 2013, from http://www.timeshighereducation.co.uk/story.asp?sectioncode=26&storycode=422290&c=1.

Goldman, M. (2012). The innovative medicines initiative: A European response to the innovation challenge. *Clinical Pharmacology & Therapeutics, 91*(3), 418–425. doi:10.1038/clpt.2011.321, doi:10.1038%2Fclpt.2011.321.

Gottweis, H., Salter, B., & Waldby, C. (2009). *The global politics of human embryonic stem cell science: Regenerative medicine in transition*. New York: Palgrave MacMillan.

Gottweis, H., & Kim, B. S. (2010). Explaining Hwang-gate: South Korean identity politics between bio-nationalism and globalization. *Science, Technology, and Human Values, 35*, 501–524.

Hogarth, S., & Salter, B. (2010). Regenerative medicine in Europe: Global competition and innovation governance. *Regenerative Medicine, 5*(6), 971–985. doi:10.2217/rme.10.81.

Huang, C., Amorim, C., Spinoglio, M., Gouvia, B., & Medina, A. (2004). Organization, programme, and structure: An analysis of the Chinese innovation policy framework. *R&D Management, 34*(4), 367–387.

International Society for Stem Cell Research. (2008). Guidelines for the clinical translation of stem cells. Retrieved January 27, 2013, from http://www.isscr.org/docs/default-source/clin-trans-guidelines/isscrglclinicaltrans.pdf.

Isasi, R. M. (2009). Policy interoperability in stem cell research: Demystifying harmonization. *Stem Cell Reviews and Reports*. doi:10.1007/s12015-009-9067-z.

Jasanoff, S. (2004). *States of knowledge: The co-production of science and social order*. London: Routledge.

Jasanoff, S. (2005). *Designs on nature: Science and democracy in Europe and the United States*. Princeton, NJ: Princeton University Press.

Johnson, P. C., Bertram, T. A., Tawil, B., & Hellman, K. B. (2010). Hurdles in tissue engineering/regenerative medicine product commercialization: A survey of North American academia and industry. *Tissue Engineering. Part A, 17*(1–2), 5–15. doi:10.1089/ten.tea.2010.0411.

Johnson, P. C., Mikos, A. G., Fisher, J. P., & Jansen, J. A. (2007). Strategic directions in tissue engineering. *Tissue Engineering, 13*(12), 2827–2837. doi:10.1089/ten.2007.0335.

Kessler, C. (2010). Retrieved March 2013 from http://www.ema.europa.eu/docs/en_GB/document_library/Presentation/2010/05/WC500090649.pdf.

Kingman, S., Yamauchi, M., Dorozynski, A., Swan, N., et al. (1992). Fetal tissue research around the world. *British Medical Journal, 304*(6827), 591–593, 597.

Lau, D., Ogbogu, U., Taylor, B., Stafinski, T., Menon, D., & Caulfield, T. (2008). Stem cell clinics online: The direct-to-consumer portrayal of stem cell medicine. *Cell Stem Cell, 3*, 591–594.

Luo, J., Flynn, J. M., Solnick, R. E., Ecklund, E. H., & Matthews, K. R. W. (2011). International stem cell collaboration: How disparate policies between the United States and the United Kingdom impact research. *PLoS One, 6*(3), e17684.

Leber, J. (2013). In the developing world, MOOCs start to get real. *Technology Review*. Retrieved March15,2013,fromhttp://www.technologyreview.com/news/512256/in-the-developing-world-moocs-start-to-get-real/.

Levine, A. (2011). Policy uncertainty and the conduct of stem cell research. *Cell Stem Cell, 8*, 132–135. doi:10.1016/j.stem.2011.01.002.

Lindvall, O., & Hyun, I. (2009). Medical innovation vs. stem cell tourism. *Science, 324*, 1664–1665.

Longaker, M., Baker, T., & Greely, H. (2007). Proposition 71 and CIRM- assessing the return on investment. *Nature Biotechnology, 25*, 513–521.

Lysaght, M., Jaklenec, A., & Deweerd, E. (2008). Great expectations: Private sector activity in tissue engineering, regenerative medicine, and stem cell therapeutics. *Tissue Engineering. Part A, 14*(2), 305–319.

Martin, P. A., Coveny, C., Kraft, A., Brown, N., & Bath, P. (2006). The commercial development of stem cell technology: Lessons from the past, strategies for the future. *Regenerative Medicine, 1*(6), 801–807.

Marchant, G., & Mossman, K. (2004). *Arbitrary and capricious: The precautionary principle in European Union Courts*. Washington, DC: American Enterprise Institute for Public Policy Research.

Mason, C., McCall, M., Culme-Seymour, E., Suthasan, S., Edwards-Parton, S., Bonfiglio, G. et al. (2012). The global cell therapy industry continues to rise during the second and third quarters of 2012. *Cell Stem Cell, 11*(6), 735–739.

McCoy, A. (2009). Biotechnology and embryonic stem cells: A comparative analysis of the laws and policies of the United States and other nations. *Loyola Law & Technology Annual, 8*(1), 63–94.

McMahon, D. S., Thorsteinsdottir, H., Singer, P. A., & Daar, A. S. (2010). Cultivating regenerative medicine innovation in China. *Regenerative Medicine, 5*(1), 35–44.

MRC. (2012). A Strategy for UK Regenerative Medicine. Document prepared by the UK Medical Research Council (MRC) in partnership with the BBSRC, EPSRC, ESRC and the Technology Strategy Board (TSB). (April).

Murray, F., & Spar, D. (2006). Bit player or powerhouse? China and stem-cell research. *The New England Journal of Medicine, 355*, 1191–1194. doi:10.1056/NEJMp068151.

National Research Council. (2009). *A new biology for the 21st century*. Washington, DC: National Academies Press.

Nerem, R., Loring, J., McDevitt, T., Palecek, S., Schaffer, D. V., & Zandstra, P. (2013). *WTEC panel report on global assessment of stem cell engineering*. Lancaster: World Technology Evaluation Center.

Normile, D. (2013). Japan to relax ban on Chimeric embryo experiments. *Science*. http://news.sciencemag.org/scienceinsider/2013/06/japan-to-relax-ban-on-chimeric-e.html.

Office of Life Sciences. (2011). Taking stock of regenerative medicine in the United Kingdom. Report of the UK Department of Health. Retrieved February 5, 2013, from https://www.gov.uk/government/uploads/system/uploads/attachment_data/file/32459/11-1056-taking-stock-of-regenerative-medicine.pdf.

Oiwa, Y. (2013). Shinya Yamanaka: The goal is still helping patients. *The Asahi Shimbun*. Retrieved March 26, 2013, from http://ajw.asahi.com/article/behind_news/people/AJ201303220009.

Pagnol, A., Rowley, E., Martin, P., & Livesey, F. (2009). Industry perceptions of barriers to commercialization of regenerative medicine products in the UK. *Regenerative Medicine, 4*(4), 549–559.

Park, M. (2012). Texas board approves rules on use of stem cells. *New York Times*. April 14, A14.

Petersen, A., & Seear, K. (2011). Technologies of hope: Techniques of the online advertising of stem cell treatments. *New Genetics and Society, 30*(4), 329–346.

Petryna, A. (2009). *When experiments travel: Clinical trials and the search for human subjects*. Princeton: Princeton University Press.

Qiu, J. (2011). China sets 2020 vision for science. *Nature, 470*, 15. doi:10.1038/470015a.

Rao, M. S. (2009). Funding translational work in cell-based therapy. *Cell Stem Cell, 9*(1), 7–10 (2011). doi:10.1016/j.stem.2011.06.009.

Regenberg, A., Hutchinson, L., Schanker, B., & Matthews, D. (2009). Medicine on the fringe: Stem cell-based interventions in advance of evidence. *Stem Cells, 27*(9), 2312–2319. doi:10.1002/stem.132.

Rosemann, A. (2013). Medical innovation and national experimental pluralism: Insights from clinical stem cell research and applications in China. *BioSocieties, 8*(1), 58–74. doi:10.1057/biosoc.2012.39.

Salter, B. (2008). Governing stem cell science in China and India: Emerging economies and the global politics of innovation. *New Genetics and Society, 27*(2), 145–159.

Salter, B. (2009a). State strategies and the global knowledge economy: The geopolitics of regenerative medicine. *Geopolitics, 14*, 1–31.

Salter, B. (2009b). China, globalisation and health biotechnology innovation: Venture capital and the adaptive state. *East Asian Science and Technology: an International Journal, 3*(4), 401–420.

Salter, B., & Salter, C. (2010). Governing innovation in the biomedicine knowledge economy: Stem cell science in the USA. *Science and Public Policy, 37*(2), 87–100.

Sinha, G. (2011). Spat over innovative medicines initiative (IMI) funding and intellectual property. *Nature Biotechnology, 29*(6), 473. doi:10.1038/nbt0611-473a, doi:10.1038%2Fnbt0611-473a.

Sleeboom-Faulkner, M. (2008). Debates on human embryonic stem cell research in Japan: Minority voices and their political amplifiers. *Science as Culture, 17*(1), 85–97. doi:org/10.1080/09505430801915455.

Sleeboom-Faulkner, M. (2013). Latent science collaboration: Strategies of bioethical capacity building in mainland China's stem cell world. *BioSocieties, 8*(1), 7–24.

Song, P. (2011). The proliferation of stem cell therapies in post-Mao China: Problematizing ethical regulation. *New Genetics and Society, 30*(2), 141–153.

Stafford, N. (2009). Germany tightens law on stem cell treatments. *British Medical Journal, 339*. doi:10.1136/bmj.b2967.

Thornley, B., Wood, D., Grace, K., & Sullivant, S. (2011). National high-tech R&D (863) program. Impact investing: A framework for policy design and analysis. Retrieved December 13, 2012, from http://www.pacificcommunityventures.org/uploads/misc/case_studies/13-High_Tech_RD_Program.pdf.

Toft, M., Philpott, D., & Shah, S. (2011). *God's Century: Resurgent religion and global politics*. New York: W.W. Norton and Company.

Turner, L. (2007). First world health care at third world prices: Globalization, bioethics and medical tourism. *BioSocieties, 2*(3), 303–326.

U.S. Department of Health and Human Services (DHHS). (2006). 2020: A new vision: A future for regenerative medicine. Retrieved January 15, 2013, from http://medicine.osu.edu/regenerative-medicine/documents/2020vision.pdf.

Xu, G. T. (2008). Establishment of stem cell bank in China. *Cell Research, 18*, 175. doi:10.1038/cr.2008.265.

Yuan, W., Sipp, S., Wang, Z., Deng, H., Pei, D., Zhou, Q. et al. (2012). Stem cell science on the rise in China. *Cell Stem Cell, 10*(1), 12–15. doi:10.1016/j.stem.2011.12.002.

Zhang, J. Y. (2012b). *The cosmopolitanization of science: Stem cell governance in China*. Basingstoke, UK and New York: Palgrave Macmillan.

Chapter 3
Policies and Practices to Enhance Multi-sectorial Collaborations and Commercialization of Regenerative Medicine

Tania Bubela, Amrita Mishra, and Debra Mathews

3.1 Introduction

Realizing the potential of stem cells will require harnessing the respective strengths of academic, biotechnology, and pharmaceutical sectors. Indeed, a complex ecosystem has formed to support the field, further including patient advocacy organizations, philanthropic foundations, and public-sector funders. Such active engagement of multiple actors is required because of the long and difficult road for cell therapies to the clinic (Daley 2010). The complexities of the translational pathways for stem cell research point to the need to bolster incentives for collaborative, multinational, multidisciplinary, and multi-sectorial research and development (Bubela et al. 2012a; Winickoff et al. 2009). However, it must also be recognized that stem cell research needs "to be incubated in academia much longer before it is ready to graduate into a business that can commercialize the technology and deliver real products" (Giebel 2005). Such incubation requires policy, funding, and infrastructure support for the pre-competitive research environment.

Intellectual property rights (IPRs) are often considered as prime incentives for technological innovation, especially in biotechnology and pharmaceutical sectors where they are seen as a shield against the risks and uncertainties of the research and development (R&D) process (Eisenberg 2003). The principal forms of IPR in the pharmaceutical sector are a blend of patents, trade secrets, and protected data—in particular, data in review with regulatory agencies during drug or device approval processes (Bubela et al. 2012a). Patents are intended to stimulate the creation of

T. Bubela (✉) • A. Mishra
School of Public Health, University of Alberta, Edmonton Clinic Health Academy, Edmonton, Alberta, Canada T6G 1C9
e-mail: tbubela@ualberta.ca

D. Mathews
Johns Hopkins Berman Institute of Bioethics, 1809 Ashland Avenue, Baltimore, MD 21205, USA

L.F. Hogle (ed.), *Regenerative Medicine Ethics: Governing Research and Knowledge Practices*, DOI 10.1007/978-1-4614-9062-3_3,

new technologies with practical applications through the grant of exclusive rights to make, use, sell, and import inventions. However, the overuse and inappropriate management of IPRs may counterintuitively hinder the pre-competitive collaborations essential for the clinical translation of stem cell research (Bubela et al. 2012a).

Collaborations are theorized to lead to the more efficient use of resources and knowledge through the reduction of negotiating costs associated with proprietary holdings, the avoidance of duplicative research within a secretive environment, and the use of standard research tools and methods, supported by repositories for data and materials. A collaborative approach to pre-competitive research is premised on the fact that therapies may take between 10 and 15 years to reach the clinic and likely longer for cell therapies (Paul et al. 2010). But patent terms extend for only 20 years plus extensions in some jurisdictions to compensate for lengthy regulatory approval processes. Further, even drugs may take between $US200 million to US$1 billion to develop, and cell therapies possibly even more, but venture capital firms are increasingly investing later in the development process, following proof of concept in humans and trending towards requiring positive Phase II clinical trial results.

The dominance of academia in translational stem cell research is evident through patent ownership, clinical trials, and the structure of the industry. In 2006, Bergman and Graff (2007a) noted that 40 % of all stem cell patents were assigned to universities or research institutions. Similarly, clinical trials, with few exceptions, are early stage and sponsored by academic institutions (Bubela et al. 2012b). The vast majority of trials remain focused on hematological cancers, while a minority (approximately 20 %) may be considered as novel and representative of the future of regenerative medicine. Such trials span cell-based therapies, the use of small molecules and biologics for regenerative goals, the use of synthetic materials, biomaterials and scaffolds, and stem cell therapy combined with gene therapy. Contrary to public expectations and the hype surrounding stem cell research, few trials use human embryonic stem cells (hESCs) and only a small number target neurological conditions (Bubela et al. 2012b).

Industry has been slow to enter the field; venture capital for the biotechnology sector has been increasingly difficult to access following the economic crisis in 2008, the emergence of business models specific to cell therapies has been slow, and regulatory pathways remain uncertain. The emerging regenerative medicine industry is dominated by small biotechnology companies, many university spin-offs, focused on tools and nontherapeutic products (48 %) and services and manufacturing (9 %) (Alliance for Regenerative Medicine (ARM) 2012). In other words, the majority of regenerative medicine firms provide research tools and reagents to support R&D efforts, which are dominated by academic institutions and, therefore, funded largely by public and philanthropic funds.

In this chapter, we explore models to facilitate multinational, multidisciplinary, and multi-sectorial collaborations in stem cell research. The need for such models has been recognized in a recent Institute of Medicine (IOM 2012) external report on the California Institute of Regenerative Medicine (CIRM), which identified disincentives to industry partnerships and a Consensus Statement by the Hinxton Group

(2012), an international consortium on stem cell ethics and law. The Consensus Statement calls for the development of formal collaborative networks for translational stem cell research. Here, our focus is on the enablement, through creative management of IPRs, of pre-competitive research with academia at its core. We start with an overview of the patent landscape in stem cell research. This allows for an understanding of claims that may block follow-on innovation and of areas where numerous overlapping claims make it difficult to define all of the IPRs within a given area and therefore to negotiate freedom to operate. The former is termed blocking patents and the latter a patent thicket or "anticommons" effect (Heller and Eisenberg 1998). We then describe emerging models in drug discovery and animal model communities that facilitate collaborations in the pre-competitive research environment. Such models range from the establishment of databases and repositories for research reagents to public–private partnerships (PPPs) that harness the capabilities and strengths of multiple sectors. The following section gives examples of such initiatives within stem cell research, mainly still focused on infrastructure support. The chapter concludes with an analysis of legal barriers to pre-competitive research collaborations, namely, restrictive licensing policies, and discusses incentive structures and best practices for facilitating collaborative, pre-competitive stem cell research.

3.2 Background: The Stem Cell Intellectual Property and Regulatory Landscape

We begin with a brief overview of the patent landscape for stem cell research, drawing on a number of published reports. This overview enables us to point to areas within the field of intense and increasing patenting activity. As explained by Bergman and Graff (2007a, b) and Winickoff et al. (2009), intense patenting may lead to broad patents by early entrants into an area, especially methods patents, that block downstream research unless licenses can be negotiated (blocking patents) and/or areas of overlapping claims that create a patent thicket. A patent thicket occurs when follow-on researchers are unable to identify or negotiate all of the intellectual property in their space, limiting their freedom to operate. Both blocking patents and patent thickets, especially in areas of research that are remote from clinical application, are detrimental to the establishment of infrastructure and collaborations that promote a pre-competitive environment.

Our own analysis of stem cell Patent Cooperation Treaty (PCT) filings indicates a steady increase in the number of patents that refer to "stem cell" in their claims from 1991 to 2007 (Mathews et al. 2013), after which there has been a downward trend in patents with "stem cell" claims (up to 2012). This trend mirrors that reported by the United Kingdom Intellectual Property Office (2010) in its *Overview of the UK National Stem Cell Network Patent Watch Landscape*. That report also points to the dominance of academic and government research institutions and researchers in patent filings. It notes that the top holder of published patent applications is Kyoto University and that of granted patents is the Wisconsin Alumni Research Foundation (WARF). The lead in the United Kingdom is the University of Edinburgh.

At the peak in 2007, Bergman and Graff (2007b) pointed to the potential problem for blocking patents combined with restrictive licensing practices to hinder the evolution of stem cell research. They noted that concerns centered on the patent estate over aspects of hESC research, spanning reagents and processes for the derivation of hESCs, culture techniques and maintenance of hESCs, manipulation and differentiation of hESCs, as well as the hESC and derivative lines themselves (Parsons et al. 2011). In 2009, Konski and Spielthenner divided the stem cell field, using patent citation and network analyses, into dominant patent clusters around hematopoietic stem cells, culturing of hematopoietic stem cells, nonhuman stem cells, embryonic-like stem cells, including neural stem cells, and finally embryonic stem cells (Konski and Spielthenner 2009). However, their analysis only began to capture the emerging field of induced pluripotent stem cells (iPSCs), billed as the ethical alternative to hESC research (Zarzeczny et al. 2009).

More recently, published studies have begun to track patenting over iPSCs, which cover methods for cellular reprogramming for pluripotency (Simon et al. 2010) and methods which link discovery, diagnostics, and therapeutics using iPSCs (Vrtovec and Vrtovec 2013). The patenting of iPSCs has not been without controversy. The first patent was granted in 2008 in Japan to Dr. Yamanaka, joint winner of the 2012 Nobel Prize in Physiology or Medicine for his work in cellular reprogramming. The patent covers the process of introducing four embryonic transcription factors into a somatic cell. Also in 2010, a California-based firm, iPierian, was granted a patent over reprogramming technology in the United Kingdom, the first outside of Japan (Russell 2010). The inventors listed on that patent were a competing group of Japanese researchers, led by Dr. Sakurada. Complicating the landscape was the grant in 2010 in the United States to Dr. Jaenisch of a patent that predated the priority dates of Yamanaka and Sakurada's patent filings. The extent to which the Jaenisch claims render the later patents obvious remains unsettled. Additionally, the Thomson 2012 patent (8,183,038) also covers a method to make iPSCs. Our own analysis of keywords in claims in PCT filings indicates that iPSC-related claims continue to increase from 2008, defying the general downward trend over that time period for other stem cell-related patents (Mathews et al. 2013).

The increase in patent filings reflects the momentum of iPSC research (Löser et al. 2012), and here lessons may be learned from prior patenting in hESC research. The key patents over hESCs were assigned in 1998 to the WARF, the technology transfer organization that operates in support of research at the University of Wisconsin–Madison (Bergman and Graff 2007a, b). At the time, limited funds were available for hESCs and so the research in Dr. Thomson's laboratory was partly funded by Geron (Menlo Park, CA, USA). In 1999, WARF granted Geron exclusive commercialization rights to six cell lineages from 5 WARF cell types. In 2001, Geron obliged WARF to exclusively license 12 additional cell types to the company, leading WARF to file a lawsuit, citing the need to be able to distribute cell lines to academic researchers. A year later, the case was settled when Geron and WARF entered into a new licensing agreement that permitted WiCell to distribute lines more freely to academic and government researchers (Bizjournal 2002). Geron was given exclusive rights to clinical use of hESC-derived neural cells, cardiac cells, and pancreatic cells, while WARF retained rights for other clinical uses. The first of the hESC patents are set to expire in 2014, while others have coverage until the late 2010s, but, only four clinical trials in clinicaltrials.gov to 2012 have used hESCs or cells derived from hESCs with no clinical application imminent. The most publicized clinical trial sponsored by Geron for spinal cord injury was halted in November 2011 after use in only four trial subjects (Simon et al. 2010). At that time, Geron discontinued its stem cell program. Advanced Cell Technology began three additional trials, two using retinal pigment epithelial cells derived from hESCs for Stargardt's macular dystrophy and one for dry age-related macular degeneration (Atala 2012). Thus for the duration of the patent terms, the WARF hESC patents were, in effect, proprietary rights over research tools. These were distributed to the research community via WiCell, a nonprofit, private laboratory setup for that purpose (Gulbrandsen 2007).

Bringing iPSCs and hESCs to the clinic will require a far greater understanding of cell biology, especially the behavior and long-term stability of iPSCs, before regulators will be willing to approve for human use. From the perspective of regulators, both hESCs and iPSCs raise significant concerns because both have been shown to be unstable in culture (Zwaka 2010; Pera 2011). The difficulties inherent in ensuring manufacturing consistency with reproducible quality of the iPSC product will add an additional regulatory burden. Indeed, the Geron trial using an hESC-derived cell product for spinal cord injury was delayed by the Food and Drug Administration (FDA) because of concerns over the formation of cysts in animal models. Geron responded with new assays for the characterization, purity, and predictability of the cell product (Bizjournal 2010). The characterization of cell products is especially important because of genetic instability of cell lines in culture, which gives rise to the dominant concern: the potential for long-term harm from tumor formation. For iPSCs that require both genetic manipulation and cell culturing, regulators are likely to require novel karyotype and genomic assays as well as additional research on the short- and long-term effects of cell line genetic instability (Goldring et al. 2011). As explained in detail by Goldring et al. (2011), the potential for tumor formation will necessitate longer-term pre-clinical studies in animal models and longer-term monitoring of clinical trial participants to assess safety. In addition, patient-specific lines will likely be considered separate products rather than universal therapeutic agents, necessitating rigorous testing for each line generated (Goldring et al. 2011). The complexity of the regulatory pathways in major markets, such as the United States and Europe, also remains uncertain and will depend on interpretation in the different jurisdictions of the extent to which iPSCs have been manipulated (Condic and Rao 2008; von Tigerstrom 2008, 2011; Zarzeczny et al. 2009). To date, only one clinical trial using iPSCs has been granted regulatory approval in Japan, where the technology originated (Russell 2010; Castelvecchi 2013).

Thus, given the barriers to clinical application of iPSCs, both research-related and regulatory, at present, the bulk of iPSC research remains far from clinical application. This raises concerns about the expanding patent landscape covering iPSCs, which sits firmly within what should remain the pre-competitive research environment. Patents granted in the near term are unlikely to be in force if or when iPSC therapies become standard of care. Instead, a complex patent environment with broad and overlapping claims over cellular reprogramming may hinder the long-term and collaborative research necessary to bring therapies to the clinic.

Patent-seeking behavior has been heavily incentivized in academic institutions since the 1980s and the passage of the United States Patent and Trademark Law Amendments Act (Bayh–Dole) that enabled US universities to hold intellectual property over federally funded inventions. Other countries, many in the absence of specific legislation, have also promoted commercialization goals for research, supported by active patenting by researchers over publicly funded research products and processes (Bubela et al. 2010; Saha et al. 2010) and the institutionalization of technology transfer (Berman 2012). Since it is unlikely that academic and other patenting activity in the stem cell field will be curtailed, promoting the pre-competitive environment requires creative management of IPRs to foster collaborative research, alongside appropriate incentives for such activities. The next section discusses some models which have arisen to foster multi-sectorial collaborations in the pre-competitive environment.

3.3 Models for Managing the Pre-competitive Environment

Stem cell research can gain insights from other biomedical research communities, namely, drug discovery and animal model communities, that are building novel innovation models for translational research (Bubela et al. 2012a; Friend 2010; Norman et al. 2011; Organization for Economic Co-operation and Development (OECD) 2011; Woodcock 2010). Such initiatives involve greater selectivity in seeking IPRs, more creative governance structures for managing intellectual property, and incentives aligned with the need to foster collaborations. These are important enabling factors in the early phases of R&D, characterized as a *pre-competitive* space, where collaborations and knowledge sharing are incentivized and competition based on IPRs is minimized. Over-reliance on IPRs in early-stage research increases both the direct costs of securing IPRs and the transaction costs of negotiating licenses to use research tools and reagents with little commercial value. Innovation costs are compounded by duplication of secretive R&D efforts; the majority of which will fail in early-stage clinical trials (Bubela et al. 2012d; Friend 2010; Munos 2006, 2009).

New models of IPR governance could be enabling factors for the pre-competitive environment. Admittedly, however, these models are relatively new, and whether they in fact improve outcomes for R&D in terms of innovative products reaching the market remains uncertain (Bubela et al. 2012d; Paul et al. 2010). The key evidence supporting such productivity derives from collaborations for global health (Moran et al. 2009, 2011). We will discuss three emerging models that are relevant to the field of stem cell research: common pool resources, open access, and PPPs. Common pool resources, also called a research commons, create the infrastructure necessary to share research tools and data to the research community from the managed environment of a biorepository/biobank or database. Management complexity may range from simple use agreements to material transfer agreements (MTAs). There is no doubt that broad access for all the sectors of R&D to research reagents and tools as well as data enhances the pre-competitive research environment, and this model is discussed more fully in the following section.

The second model is open access whereby norms within a research community are crafted to facilitate access to knowledge as well as collaboration. Members of the community, including funders and collaborators, create rules on how knowledge is produced and used and then enforce those rules. The open access model does not necessarily rely on IPRs, and a leading example is the Structural Genomics Consortium (SGC). Formed in 2004, the SGC is an international (mainly United Kingdom and Canada) nonprofit scientific research consortium funded by governments, foundations, and industry working in drug discovery (Edwards et al. 2009, 2011; Lee et al. 2009; Weigelt 2009). Large pharmaceutical companies such as GlaxoSmithKline (GSK) and Pfizer support the SGC. The SGC was initially aimed at identifying three-dimensional structures of proteins, in which regard it has been highly successful, depositing may such structures to public databases and producing high-impact publications (Gileadi et al. 2007). However, in recognition of the fact that drug discovery is accelerated through the availability of research tools, the SGC has expanded its initiative to make publicly available probes for epigenetics research (Edwards et al. 2009, 2011). Under its open access policy, the SGC does not seek, nor permit its affiliated public or private sector researchers and collaborators to seek, patents that would grant exclusive rights over its research outputs. Further, its leadership has taken a strong advocacy role in supporting pre-competitive research in an environment that foregoes patent rights. As stated by Aled Edwards of the SGC: "The fundamental problem is that industry collectively focuses too many resources on proof-of-concept studies for too few targets, and the studies are done in a proprietary way, with little collective learning. Further, because one 'secret' failure in proof of concept is never enough to dissuade others, these studies encumber the limited resources of industry for years, thereby limiting the ability of industry to pursue new and potentially relevant drug targets" (Edwards et al. 2009).

The third model comprises open innovation strategies that range from information databases to facilitate knowledge exchange about or trade in IPRs (e.g., patent pools and clearinghouses) to the facilitation of joint research partnerships through exchange programs, research alliances, or PPPs. One of the best known examples of the latter is the Innovative Medicines Initiative (IMI). IMI is designed to accelerate drug development and is a joint undertaking between the European Union and the European Federation of Pharmaceutical Industries and Associations (Goldman 2011). As a funder of research consortia comprising academic, biotech, and pharmaceutical sectors, IMI enforces an IP strategy that facilitates the confidential sharing of data, materials, and other IP amongst members within funded consortia. Each consortium must then develop a commercialization plan over any developed IPRs, addressing priority, ownership, and licensing terms, including the allocation of royalty revenues generated.

Despite policies aimed at commercialization, IMI continues to prioritize traditional dissemination through publication and data accessibility. Most importantly, however, the focus is on collaboration, and in the first 2013 independent evaluation of IMI, there was a positive correlation between collaborative networking and the quantity and quality of publication (IMI Executive Office 2013). For 2010–2011, the average citation impact for IMI project papers was 1.55 (compared to a world average of 1.0). Publication output also increased substantially from 2010 to 2012, with a doubling in the 2011–2012 period, although output was, not surprisingly, higher for academic researchers compared to collaborators from industry and biotechnology companies. Encouragingly, about 40 % of all publications by IMI researchers were cross-sector, for example, between academic institutions and small medium enterprises (SMEs). The report also highlighted the increased output of researchers associated with patient organizations in top-quartile journals (IMI Executive Office 2013). It is too early to tell, however, whether this enhanced research output will result in more innovative products, diagnostic tests, and methods in practical use in clinical settings and the broader marketplace.

3.4 Pre-competitive Models in Stem Cell Research

Pre-competitive models are also gaining traction within stem cell research, primarily in the development of repositories for research reagents and data. As stated in the previous section, significant experience in the distribution of research reagents exists within the animal model community, for example, the mouse model for human disease research community (Bubela et al. 2012d). Mouse repositories may distribute live mice, gametes, or mouse embryonic stem cells. The most established and largest mouse repository, Jackson Laboratory (JAX), facilitates access to mouse models through standard agreements with donors of new mouse strains. Under the donor agreement, JAX distributes patented and unpatented mice to academic and not-for-profit researchers using a simple notification that the mice be used for research purposes and may not be sold or transferred to third parties without permission. However, mice are distributed to industry or for commercial use only if permitted by the donor. In that event, JAX acts as a broker, distributing the mice only once a license agreement has been negotiated between the donor and industry recipient (Einhorn and Heimes 2009). In 2009, leaders in the mouse model community, including directors of repositories, drafted the "Rome Agenda," which recommends "materials and data be shared under the least restrictive terms possible" (Schofield et al. 2009). The Agenda calls for the promotion of a "mouse research commons" for mouse resources and data to further pre-competitive research and our understanding of gene function (Voell et al. 2010; Schofield et al. 2009; Hancock et al. 2008). Einhorn and Heimes (2009) describe JAX as already promoting the mouse academic research commons.

Stem cell banks, for example, the former US National Stem Cell Bank (NSCB), continuing as an hESC and iPSC bank (the Wisconsin International Stem Cell (WISC) Bank) and the United Kingdom Stem Cell Bank (UKSCB), are similar to mouse repositories in that they store, expand, manipulate, and distribute embryonic and adult cell lines. Currently, however, lack of harmonized distribution policies and standards reduces the interoperability of cell banks and limits their transnational accessibility and utility. Furthermore, there are many private banks held in research institutions and elsewhere, which hold lines that are not broadly available to outside investigators, and not necessarily derived and characterized according to agreed-upon standards (Laursen 2009).

As a move towards harmonization, the International Stem Cell Banking Initiative (ISCBI) adopted a 2008 Consensus Guidance for Banking and Supply of Human Embryonic Stem Cell Lines for Research Purposes (ISCBI 2009). This document provides technical and ethical guidelines for procuring, documenting, banking, and distributing hESCs for research use. Efforts to catalogue and disseminate standardized cell-line information have been undertaken by the European Human Embryonic Stem Cell Registry (hESCreg), the International Society for Stem Cell Research (ISSCR), and the International Stem Cell Registry (ISCR) (Knoppers and Isasi 2010). The development of well-networked national repositories with staff trained in cell-line propagation, registry maintenance, and distribution could provide the material base upon which to apply uniform standards. However, as has been noted in the mouse model community (Schofield et al. 2010), efforts also need supervisory bodies with the capacity and authority to enforce guidelines across facilities. The extent of centralization that is feasible is open to debate. At the national level, centralized facilities would save on the costs of maintaining cell banks at multiple locations and adequate networking could enable oversight of specimen donation, storage, and use (Winickoff 2006). However, banking facilities with transnational operations may face challenges due to policy differences with regard to human research subjects protections and other regulatory requirements.

The ISSCR and the ISCBI are also attempting to address the policy challenges emerging in the use of cord blood-derived stem cells and iPSCs (Knoppers and Isasi 2010). The use of iPSCs also faces regulatory and ethical challenges, owing to donor anonymity (Mathews et al. 2011; Resnik this volume) and reuse of iPSCs in controversial forms such as iPSC-derived gametes (Zarzeczny et al. 2009; Caulfield et al. 2010). The research use of iPSCs, therefore, necessitates a nuanced approach to informed consent because specimen donors may need to be traced and recontacted for their consent for uses unknown at the time of initial consent (Lowenthal et al. 2012). Donors' views and expectations of research on their specimens are also conditioned by the prospect of commercial application. Donors may have varying reactions, from outright withdrawal to expectation of financial benefit from commercial proceeds (Zarzeczny et al. 2009), yet the most useful consents will likely be those where donors have agreed to use for commercial purposes. Such consent issues complicate the distribution of cell lines, requiring more complex terms of use within MTAs than necessary for nonhuman lines, such as mouse embryonic stem cells.

Nevertheless, efforts at harmonization, if successful, should facilitate access to these research materials. Indeed, CIRM has funded a human pluripotent stem cell (hPSC)/iPSC repository (http://www.cirm.ca.gov/our-funding/research-rfas/hpsc-repository). The United Kingdom has launched a national iPSC bank, through the Human Induced Pluripotent Stem Cell Initiative (HIPSCI), funded by the Medical Research Council and the Wellcome Trust (http://www.hipsci.org). IMI is funding an academic-industry consortium of 10 pharmaceutical companies and 23 academic groups to establish StemBANCC, a repository that will make available 1,500 iPSC lines (http://www.imi.europa.eu/content/stembancc). It is also set to fund a European iPSC repository comparable to that funded by CIRM (http://www.imi.europa.eu/sites/default/files/uploads/documents/8th_Call/IMI_8thCallText_FINAL.pdf). While there has been no mention of collaboration amongst these three iPSC repositories, their development will facilitate the sharing of common lines. Many smaller banks also exist, such as the RIKEN Bioresource Center Cell Bank (http://www.brc.riken.go.jp/lab/cell/english/), though how they will coordinate, or not, with the larger banks remains to be seen.

With respect to the model of information clearinghouses, The Hinxton Group Consensus Statement (Hinxton Group 2012) recommended the creation of networked resources or hubs for sharing technical, provenance, and IPRs information about available stem cell lines. While such sharing is anticipated to facilitate progress in stem cell science through increased access to data and materials, a potential barrier is made quite real with the increasing influence of Japan. Japanese institutions, Kyoto University and CiRA, have become a dominant presence in iPSC-related IPRs, which complicates the ability of information sharing due to language and translation issues (Hinxton Group 2012). Translation is time consuming and expensive. Furthermore, national regulations in both Japan and China make it difficult to share cell lines derived from their citizens' tissues across national boundaries, as such tissues are considered a national bioresource. These restrictions thus prohibit, for example, the deposition of Japanese iPSC lines in the UKSCB (Hinxton Group 2012). Differences in regulatory requirements for research with human and clinical trial approval also pose challenges for developing international standards for cell line banks and registries.

In addition to biobanks, databases, and clearinghouses, an even more innovative approach at providing infrastructure to facilitate translational stem cell research is being led by the National Heart, Lung and Blood Institute (NHLBI). The goal of this program, the Production Assistance for Cellular Therapies (PACT), is to provide clinical investigators with access to production facilities, training, and regulatory assistance (Reed et al. 2009). PACT consists of three large groups, involving multi-institutional collaboration of private and public entities: the administrative center, the EMMES Corporation (a Contract Research Organization (CRO)); five facilities that are compliant with current Good Manufacturing Practice (cGMP); and a steering committee with representatives from each facility, EMMES and NHLBI and external NHLBI appointed chairs. The five cGMP facilities are located throughout the United States at Baylor College of Medicine Center for Cell and Gene Therapy; Beckman Research Institute of the

City of Hope Center for Applied Technology Development; Center for Human Cell Therapy, Boston; University of Minnesota Molecular and Cellular Therapeutics Facility; and Waisman BioManufacturing, housed at the University of Wisconsin–Madison. The significance of PACT is that it provides much-needed assistance with scale-up protocols needed for clinical trials. Additionally, PACT offers training through web-based seminars and workshops, focusing on GMP production techniques and regulatory issues. PACT supports researchers through preclinical and Phase I trials of therapies including regulatory T-lymphocyte cells, natural killer cells, adipose-derived stem cells, cardiac progenitor cells for cardiac disease, hematopoietic progenitor cells (HPCs) for central nervous system applications, cytotoxic T lymphocytes, and dendritic cells.

In conclusion, innovative approaches are being developed to facilitate precompetitive stem cell research. Many are being developed in parallel to initiatives in more mature fields such as genomics and animal model research. However, others, such as the PACT, address issues specific to stem cell research. The end point should be to facilitate access to knowledge and research tools required to move the field forward. The next section, therefore, addresses some of the legal and practical barriers to the flow of knowledge, research tools, and reagents.

3.5 Overcoming Material Transfer Agreements and Licensing as Barriers to Collaborative Models

Access to research tools and data may be limited by restrictive licensing terms and MTAs—a type of license agreement that set terms of use and access to research reagents, such as cell lines. MTAs establish important legal relationships between academia, industry, and entrepreneurs in private sector firms (Mirowski 2008; Mowery and Ziedonis 2007). They are permissions to use proprietary or nonproprietary materials that are in the control of the provider and come with contractual rights and obligations (Rodriguez 2008). Although MTAs ostensibly facilitate collaborative exchange, they can significantly delay access to resources and increase transaction costs, since their content ranges from simple conditions of use to complex terms (Streitz and Bennett 2003). In addition, while patents are legally, geographically, and temporally defined and bounded, MTAs have the freedom associated with contractual formulations to be much more wide-ranging in setting restrictions. This is an important distinction that for repositories could be the difference between losing, breaking even, or being profitable on cell distribution.

MTAs usually contain provisions regarding the physical handling, use, and further distribution of the material (Winickoff et al. 2009). Often they limit onwards distribution to third-party researchers. Commonly, MTAs contain standard terms that limit liability for the provider, provide standard disclaimers as to the quality of the material provided, and set dispute resolution mechanisms and legal jurisdiction, as well as timelines for the relationship between provider and user. However, MTAs may also restrict the use of materials by recipients in a manner that does not facilitate activities within the pre-competitive research environment. For example, MTAs may contain clauses that limit an academic researcher's ability to publish in the scientific literature, through delays or complete restrictions. Further, MTAs may contain reach-through rights that give a provider rights to royalties, future licensing options, or other advantages from modifications or inventions made using the transferred material. These clauses can be an obstacle in transfers and are discouraged in most best-practice guidelines for the licensing of research tools, including practice guidelines for resources generated using National Institutes of Health (NIH) funding (Association of University Technology Managers (AUTM) 2007; NIH 1999; OECD Guidelines for the Licensing of Genetic Inventions 2006). Reach-through rights are philosophically problematic because they extend proprietary rights far beyond those that are granted by patent claims (Rai and Eisenberg 2004).

The drafting and negotiation of MTAs is handled by institutional legal counsel often located within technology transfer offices (TTOs) or research services offices. TTOs and/or university–industry liaison offices manage research partnerships, sponsored research, and commercialization activities such as patenting, licensing of technologies, and creating spin-off companies. These offices serve innovation systems by linking researchers, industry, and healthcare organizations (Miller et al. 2009). In finalizing MTAs, these offices work with variable university and funding agency policies as well as researcher practices.

Delays in negotiating MTAs are seen by biomedical researchers in many fields to be significant barriers to research and timely access to reagents (Bubela et al. 2012c, d). Such effects have been noted in hESC research, where access to lines was severely restricted due to President Bush's policy limiting federal funding for hESC research to preexisting lines in 2001 (Rao 2006). McCormick et al. (2009) used MTAs to track the shipment of 1662 vials of stem cells distributed from two US repositories, WiCell (743 vials between 2000 and 2007) and the Harvard Stem Cell Institute (HSCI) (919 vials between 2004 and 2007). The strategies of the two repositories differed in that WiCell executed one MTA for every vial shipped, while HSCI issued multiple lines under one agreement. In addition, until fall 2005, HSCI users were charged only for shipping, whereas each WiCell line costs US$ 5,000. These differences resulted in HSCI distributing "nearly 200 more vials than WiCell in about half the time" (McCormick et al. 2009). In 2005, WiCell dramatically cut its fees, and, as explained by Gulbrandsen (2007), WARF and WiCell simplified their licensing and material sharing policies in response to complaints from administrators and scientists at major research universities. First, WARF relieved industry sponsors of the obligation to seek a commercial license with WARF, in a move to encourage private sector funding of hESC research. Second, WARF and WiCell waived fees on inter-researcher sharing of hESCs and asked for an ethical-use MTA only for transfers of Wisconsin hESCs (Gulbrandsen 2007).

However, the story is not that simple. In the context of public controversy and sensitivity to ethical concerns, MTAs can be an essential way for the provider to ensure the cells are used in an ethical manner consistent with the donor's intent. For the WARF lines, unique MTAs were required so that specific guarantees regarding the use of cells from embryos could be included to honor agreements made in informed consent documents signed by embryo donors. For example, each hESC line derived by Dr. Thomson's laboratory at the University of Wisconsin had its own MTA clearly delineating acceptable uses of the material as per the donors' consent for the use of their embryos. Once materials such as cell lines are shared with other groups, it becomes difficult to ensure that use would conform to the donors' consent reflected in the initial WARF MTA. For other lines, donors consented to specific uses of their cells from their embryos. For example, some of the Melton lines, distributed by HSCI and generated by Dr. Melton, can only be used for diabetes research, because of the original donor consent. Even though the lines may work well for a variety of research, they cannot be used for other purposes. An MTA is therefore the only practical way to restrict other uses and honor the intentions of the donors. Lineage-specific restrictions can thus lead to conflicts, as researchers may be unable to use the line in a manner consistent with their experimental intentions. Nevertheless, MTAs remain the strongest legal means through which to enforce use with respect to donor consent.

Where donor consent for use of materials is not an issue, however, MTAs may be problematic, and lessons may be learned from other research communities. In the mouse research community, the OncoMouse and Cre–Lox controversies spurred universities and funders to frame nonexclusive licensing policies to promote widespread use of research tools and reagents (Bubela et al. 2012d; Hanahan et al. 2007; Murray 2010; Rossant and McMahon 1999). A nonexclusive licensing policy does not preclude the use of MTAs to ensure donor consent, but instead is a policy directed towards the broad dissemination of research tools. The first lesson in this regard comes from the OncoMouse, genetically modified to have a predisposition towards cancer. The OncoMouse was developed and patented by researchers at Harvard University and licensed exclusively to DuPont, which provided the research funding. Controversy arose initially from the broad patent covering the OncoMouse technology and later in DuPont's restrictive licensing policies. DuPont made no effort to generate improved OncoMouse models or to use the mouse in its own research. Instead, it focused on generating revenues through high-cost sublicensing to Pharma for modification and use of the mice for preclinical testing of anticancer therapies.

In terms of access to academic researchers, DuPont imposed restrictive terms contrary to community norms of the mouse community and distributed the mice via the supplier Charles River Laboratories. These terms included limits on the informal exchange of mice amongst researchers, including novel lines made by the researchers themselves; annual disclosure of published and unpublished findings using the mice to DuPont; and reach-through rights on future discoveries made using the OncoMouse. The response from the scientific community to these onerous terms was overwhelmingly negative, with researchers ignoring the patent claims or resisting through institutional channels. Institutional negotiation was significant in altering the initially restrictive conditions of use set by DuPont. After 4 years of negotiation, DuPont, JAX Labs, and the NIH signed a memorandum of understanding (MOU) that allowed researchers to exchange OncoMice for nonprofit purposes, using an agreement with simple conditions of use and without reporting requirements and reach-through rights. The MOU enabled JAX and other public repositories to distribute. OncoMouse lines, making them widely accessible to the academic research community (Bubela et al. 2012d).

The case of *Cre–Lox* mice is comparable to OncoMouse. While OncoMouse originated in a Harvard laboratory, the powerful *Cre–Lox* technology for understanding gene function had its origins in the life sciences division of DuPont. The technology enables specific genes to be turned on or off at differing developmental stages or in specific tissues. DuPont strictly controlled access to *Cre–Lox* mice using restrictive terms such as reach-through rights to inventions made using the technology, monitoring of researchers and institutions, and exclusion of non-signatories from access to materials and model organisms. This situation was lifted when an MOU was negotiated between DuPont, the NIH, the University of California (UC), and JAX. The MOU allowed JAX or universities to distribute and share *Cre–Lox* mice with a simple, standardized conditions of use agreement that eliminated reach-through licensing. The positive effect of these two cases included a 1999 NIH policy discouraging the patenting of mouse tools developed via NIH intramural funding (Schofield et al. 2009). In an excellent empirical study, Murray et al. (2009) demonstrated that openness in OncoMouse and *Cre–Lox* technology following their respective MOUs not only encouraged a greater quantity of follow-on research measured through citation counts, but also enabled new researchers to enter the field. This study clearly demonstrated the positive impact of developing institutional mechanisms to develop and distribute research tools (Bubela et al. 2012d).

Many guidelines now exist for best practices in licensing of technologies and research reagents developed in publicly funded research laboratories (e.g., OECD 2006). One practice is the implementation of standardized MTAs with simplified, nonrestrictive terms of transfer, such as the NIH's Uniform Biological Material Transfer Agreement (UBMTA), which provides a standard format to simplify negotiation of terms and expedite dissemination of resources (NIH 1995). However, standard form agreements are difficult to implement in practice and have been resisted for dissemination of biological materials compared to open access to data and deposit of data in public databases. In the absence of such standard agreements, institutions should consider the following in negotiating MTAs (Gold and Bubela 2007). There needs to be a realistic assessment of value of the material and timelines for commercial or clinical application. The complexity of the agreement should reflect the goal, especially when that goal is broad distribution of research reagents.

Whenever possible, technologies should be licensed nonexclusively, with the recognition that sometimes exclusive licenses are unavoidable, for example, in the generation of a spin-off company. In either event, research institutions should retain rights to utilize the material for noncommercial research purposes, and, if possible, that right should extend to all such research in publicly funded institutions. Another key point is to restrict benefits, including royalties and research use, to the material being licensed and not reach through to inventions made using the licensed material. In other words, if reach-through rights are frowned upon for the private sector, the same should apply to public sector, even if those research rights are intended to expand noncommercial research access to follow-on innovations developed in either the private or public sectors. Finally, as discussed above with respect to MTAs for stem cell lines, in fields that tie materials to patient records or donor consent, MTAs will, of necessity, be more complex than those that do not need to consider donor consent.

3.6 Incentivizing Investment in Pre-competitive, Collaborative Models

A final issue that confronts policy makers and funders in incentivizing collaborative research is overcoming cultural barriers to the sharing of data and reagents. Incentivizing sharing of data and reagents as well as collaboration is complicated by strongly entrenched incentives towards commercializable research outputs (Bubela et al. 2010; Caulfield et al. 2012). Universities and other research institutions are under dual pressures from funders towards a market orientation as well as knowledge generation as a public good (Berman 2012; Caulfield et al. 2012). The result is conflicting attitudes and policies from funders, governments, and institutions on the value of commercialization, industry linkages, and IPRs versus more basic or blue-sky research (Caulfield et al. 2012). While commercialization activities are encouraged as one mechanism for ensuring a return on public sector investment in research, such activities are also associated with secrecy, a reduction in collaborative research activities, and the withholding of materials and data (Bubela et al. 2010; Walsh et al. 2007; Walsh and Hong 2003). The latter is contrary to building a pre-competitive research environment facilitative of downstream clinical application.

Here we have argued that clinical translation of stem cell research will, in fact, require a far greater understanding of biological systems, an enhanced pre-competitive environment, and greater collaborative engagement from multiple sectors. It is therefore essential to move away from simplistic metrics that capture a linear commercialization model of innovation, through input–output measures along an innovation pipeline (Langford et al. 2006). As stated by John H. Lingle, performance metrics are key drivers of behavior, but "you get what you measure. Measure the wrong thing and you get the wrong behaviors."

Simple metrics are commonly used to judge the performance of TTOs and institutions in commercialization activities and include quantitative variables of invention disclosures, patents filed and granted, spin-off companies created, and licensing revenues. Such simple count data, however, do not reflect the highly networked and iterative processes needed for successful translational science (Bubela et al. 2010, 2012a). Other data may augment simple counts in an attempt to add a measure of quality. For example, publication quality may be measured through Journal Impact Factors and citation counts; the quality of individual researchers may be measured by their H-index or similar index. Similar measures of quality exist for patents where correlations exist between number of claims, number of jurisdictions covered by the patent family, and the number of times a patent is cited (OECD 2009;

Lanjouw and Schankerman 2004). Importantly, the number of citations to a patent is related to its economic value (Trajtenberg 1990; Hall et al. 2001, 2005) as is the degree to which a patent has been litigated (Allison et al. 2010).

However, while measures of quality are an improvement of simple count and economic data, such metrics do not focus on the key aim of fostering multi-sectorial collaboration to enhance the pre-competitive research environment. Towards that goal, new scientometric and bibliometric tools enable the capture and analysis of increasingly collaborative translational research pathways (Bubela et al. 2010) through linkages between researchers (co-publication or co-inventorship) and institutions (author/inventor affiliations and inventor–assignee relationships). Network analyses can visualize not only the collaborative projects but also position those projects within the broader publication network of their respective fields. For example, Bubela et al. (2010) used network analyses and visualizations to describe the extent to which researchers within the Canadian Stem Cell Network collaborated with other researchers and how those collaborations were positioned within the international field. Tracked over time, network analyses can identify emerging or persistent collaborative relationships.

Network analyses are made up of nodes (individuals, objects, institutions, firms, or organizations) and the links between them. Questions may be asked about the character and strength of the collaborations, especially the diversity of actors required for translational research efforts (Huzair et al. 2011). Network statistics can build evidence for the strength of the collaborations and their outputs. Statistics for network nodes include number of distinct collaborators, characteristics of collaborators (e.g., type of individual, institutional type, H-index, seniority, field, education, employment, country/region), and directionality of the collaborative links (unidirectional, bidirectional, weighted, or unweighted). Network analyses can identify central leaders in fields, brokers, gatekeepers, and activity hubs. In addition, network level measures include Density (total number of ties or links divided by the total number of possible links) and Centrality (identifies lead individuals or organizations in a network). In other words, the use by funding agencies and institutions of metrics and analyses focused on networks and collaborations and the quality of outputs can be a key incentive to drive such behavior. Many funding agencies, with IMI as a prime example, now focus on large-scale, multi-sectorial collaborations for targeted research funding, a significant driver for such behavior (IMI Executive Office 2013).

In addition, the policies that promote contributions to data and biological repositories also facilitate the expansion of the pre-competitive environment. However, while a myriad of policies for data and materials exist, largely promoted by public funders, these will have limited impact on individual researchers and the research environment without some attention paid to enforcement (Schofield et al. 2010). Enforcement points may be at the institutional level, through evaluation of researcher performance, tenure, and promotion; at the funder level through withholding of research funds or application review without proof of data/materials submission to a repository; or by journals requiring evidence of such submission prior to publication. The latter is already an effective incentive for the deposit of sequence data into

GenBank, the NIH's annotated collection of all publicly available DNA sequences for about 260,000 species (Benson et al. 2013).

For biobanks and databases, the main pre-competitive model operational within the stem cell research field, these repositories can incentivize contributions by simplifying deposit mechanisms, tracking submission sources, requiring acknowledgement of the originator of the material or data in any research, or by creating an identifier for material or data through which use may be tracked and reported to the originator in the same way that publication citation statistics are reported (Schofield et al. 2010). Such measures are under development in the animal model community. Institutions and funders should encourage such metrics as an indication of the broader utility of the research output.

3.7 Conclusion

Translational stem cell research requires both strong incentives for collaboration and the expansion of the pre-competitive environment wherein knowledge flows are enhanced by enabling the circulation of research tools, data, and platform technologies. This requires careful consideration of how IPRs are sought and how they are managed, and with consideration given to the investment, both public and private, needed to incentivize the development of new therapies that will hopefully flow from the pre-competitive research environment. The translational stem cell field has much to learn from other more established areas of research, such as the animal model and genomics communities. Institutions, funders, and governments need to implement appropriate metrics and enforcement structures, directed at quality of outputs and collaborative activity. However, creative metrics that incentivize collaborations remain difficult to implement within conservative and traditionally siloed academic reward systems.

References

Alliance for Regenerative Medicine. (2012). *Annual industry report*. Retrieved March 11, 2013, from http://alliancerm.org/news/download-must-have-industry-report.

Allison, J. R., Lemley, M. A., & Walker, J. (2010). Patent quality and settlement among repeat patent litigants. *Georgetown Law Journal, 99*, 677–712.

Association of University Technology Managers. (2007). *In the public interest: Nine points to consider in Licensing University Technology*. Retrieved March 18, 2013, from https://www.autm.net/Nine_Points_to_Consider.htm.

Atala, A. (2012). Human embryonic stem cells: Early hints on safety and efficacy. *Lancet, 379*(9817), 689–690.

Benson, D. A., Cavanaugh, M., Clark, K., Karsch-Mizrachi, I., Lipman, D. J., Ostell, J., et al. (2013). GenBank. *Nucleic Acids Research, 41*(Database issue), D36–D42.

Bergman, K., & Graff, G. D. (2007a). *Collaborative IP management for stem cell research and development*. Retrieved March 11, 2013, from http://hinxtongroup.files.wordpress.com/2010/10/bergman_cip-pipra_20071.pdf.

Bergman, K., & Graff, G. D. (2007). The global stem cell patent landscape: Implications for efficient technology transfer and commercial development. *Nature Biotechnology, 25*(4), 419–424.

Berman, E. P. (2012). *Creating the market university: How academic science became an economic engine*. Princeton, NJ: Princeton University Press.

Bizjournal. (2002). *WARF, Geron reach agreement on stem cells*. January 9. Retrieved March 14, 2013, from http://www.bizjournals.com/milwaukee/stories/2002/01/07/daily22.html?page=all.

Bizjournal. (2010). *Geron stem cell trial for spinal cord injury gets FDA approval*. 30 July. Retrieved March 11, 2013, from http://www.bizjournals.com/sanjose/stories/2010/07/26/daily103.html.

Bubela, T., FitzGerald, G. A., & Gold, E. R. (2012a). Recalibrating intellectual property rights to enhance translational research collaborations. *Science Translational Medicine, 4*(122), 122cm123.

Bubela, T., Li, M. D., Hafez, M., Bieber, M., & Atkins, H. (2012b). Is belief larger than fact: Expectations, optimism and reality for translational stem cell research. *BMC Medicine, 10*(133), 1741–7015.

Bubela, T., Reshef, A., Li, M. D., Atkins, A., Caulfield, T., Culme-Seymour, E., et al. (2012c). Enhancing advanced cell therapies (EnACT): Invitation to an online forum on resolving barriers to clinical translation. *Regenerative Medicine, 7*(6), 735–740.

Bubela, T., Schofield, P. N., Ryan, C. D., Adams, R., & Einhorn, D. (2012d). Managing intellectual property to promote pre-competitive research: The mouse as a model for constructing a robust research commons. *Journal of Law, Information and Science, 22*(1), 98–121.

Bubela, T., Strotmann, A., Adams, R., & Morrison, S. (2010). Commercialization and collaboration: Competing policies in publicly funded stem cell research? *Cell Stem Cell, 7*(1), 25–30.

Castelvecchi, D. (2013) Embryo-like stem cells enter first human trial. *Nature News Blog*. Retrieved February 14, 2013, from http://blogs.nature.com/news/2013/02/embryo-like-stem-cells-enter-first-human-trial.html.

Caulfield, T., Harmon, S. H., & Joly, Y. (2012). Open science versus commercialization: A modern research conflict? *Genome Medicine, 4*(2), 17.

Caulfield, T., Scott, C., Hyun, I., Lovell-Badge, R., Kato, K., & Zarzeczny, A. (2010). Stem cell research policy and iPS cells. *Nature Methods, 7*(1), 28–33.

Condic, M. L., & Rao, M. (2008). Regulatory issues for personalized pluripotent cells. *Stem Cells, 26*(11), 2753–2758.

Daley, G. Q. (2010). Stem cells: Roadmap to the clinic. *The Journal of Clinical Investigation, 120*(1), 8–10.

Edwards, A. M., Bountra, C., Kerr, D. J., & Willson, T. M. (2009). Open access chemical and clinical probes to support drug discovery. *Nature Chemical Biology, 5*(7), 436–440.

Edwards, A. M., Isserlin, R., Bader, G. D., Frye, S. V., Willson, T. M., & Yu, F. H. (2011). Too many roads not taken. *Nature, 470*(7333), 163–165.

Einhorn, D., & Heimes, R. (2009). Creating a mouse academic research commons. *Nature Biotechnology, 27*(10), 890–891.

Eisenberg, R. (2003). Patenting genome research tools and the law. *Comptes Rendus Biologies, 326*(10–11), 1115–1120.

Friend, S. H. (2010). The need for precompetitive integrative bionetwork disease model building. *Clinical Pharmacology & Therapeutics, 87*(5), 536–539.

Giebel, L. B. (2005). Stem cells—A hard sell to investors. *Nature Biotechnology, 23*(7), 798–800.

Gileadi, O., Knapp, S., Lee, W. H., Marsden, B. D., Muller, S., Niesen, F. H., et al. (2007). The scientific impact of the structural genomics consortium: A protein family and ligand-centered approach to medically-relevant human proteins. *Journal of Structural and Functional Genomics, 8*(2–3), 107–119.

Gold, E. R., & Bubela, T. (2007). Drafting effective collaborative research agreements and related contracts. In A. Krattiger, R. T. Mahoney, L. Nelsen, et al. (Eds.), *Intellectual property management in health and agricultural innovation: A handbook of best practices*. Oxford, UK: MIHR and Davis; USA: PIPRA. http://www.ipHandbook.org. Accessed 11 March 2013.

Goldman, M. (2011). Reflections on the innovative medicines initiative. *Nature Reviews Drug Discovery, 10*(5), 321–322.

Goldring, C. E. P., Duffy, P. A., Benvenisty, N., et al. (2011). Assessing the safety of stem cell therapeutics. *Cell Stem Cell, 8*, 618–628.

Gulbrandsen, C. (2007). WARF's licensing policy for ES cell lines. *Nature Biotechnology, 25*(4), 387–388.

Hall, B. H., Jaffe, A. B., & Trajtenberg, M. (2001). *The NBER patent citation data file: Lessons, insights and methodological tools*. NBER Working Paper 8498. Retrieved March 11, 2013, from http://www.nber.org/papers/w8498.

Hall, B. H., Jaffe, A., & Trajtenberg, M. (2005). Market value and patent citations. *The RAND Journal of Economics, 36*(1), 16–38.

Hanahan, D., Wagner, E. F., & Palmiter, R. D. (2007). The origins of oncomice: A history of the first transgenic mice genetically engineered to develop cancer. *Genes and Development, 21*(18), 2258–2270.

Hancock, J. M., Schofield, P. N., Chandras, C., Zouberakis, M., Aidinis, V., Smedley, D., et al. (2008). CASIMIR: Coordination and sustainability of international mouse informatics resources. In *Athens, 2008 (8th IEEE International Conference on BioInformatics and BioEngineering, BIBE 2008)*. Retrieved March 11, 2013, from http://www.casimir.org.uk/storyfiles/58.0.hancocketal-ieeeathens.pdf.

Heller, M. A., & Eisenberg, R. S. (1998). Can patents deter innovation? The anticommons in biomedical research. *Science, 280*(5364), 698–701.

Hinxton Group. (2012). *Consensus statement on data and materials sharing and intellectual property in pluripotent stem cell science in Japan and China*. Retrieved March 11, 2013, from http://www.hinxtongroup.org/consensus_hg12_final.pdf.

Huzair, F., Borda-Rodriguez, A., & Upton, M. (2011). Twenty-first century vaccinomics innovation systems: Capacity building in the global south and the role of product development partnerships (PDPs). *OMICS A Journal of Integrative Biology, 15*(9), 539–543.

Innovative Medicines Initiative Executive Office. (2013). *Bibliometric analysis of ongoing projects: Innovative medicines initiative joint undertaking*. Retrieved March 15, 2013, from http://www.imi.europa.eu/sites/default/files/uploads/documents/IMI_BibliometricReport_2_final.pdf.

Institute of Medicine. (2012). *The California institute for regenerative medicine: Science, governance, and the pursuit of cures*. Consensus Report. Retrieved March 11, 2013, from http://www.iom.edu/Reports/2012/The-California-Institute-for-Regenerative-Medicine-Science-Governance-and-the-Pursuit-of-Cures.aspx.

International Stem Cell Banking Initiative. (2009). Consensus guidance for banking and supply of human embryonic stem cell lines for research purposes. *Stem Cell Reviews and Reports, 5*(1558–6804 (Electronic)), 301–314.

Knoppers, B. M., & Isasi, R. (2010). Stem cell banking: Between traceability and identifiability. *Genome Medicine, 2*(10), 73.

Konski, A. F., & Spielthenner, D. J. F. (2009). Stem cell patents: A landscape analysis. *Nature Biotechnology, 27*(8), 722–726.

Langford, C. H., Hall, J., Josty, P., Matos, S., & Jacobson, A. (2006). Indicators and outcomes of Canadian university research: Proxies becoming goals? *Research Policy, 35*(10), 1586–1598.

Lanjouw, J. O., & Schankerman, M. (2004). Patent quality and research productivity: Measuring innovation with multiple indicators. *The Economic Journal, 114*(495), 441–465.

Laursen, L. (2009). (Stem cell) banking crisis. *Nature Reports Stem Cells*. doi:10.1038/stemcells.2009.122.

Lee, W. H., Atienza-Herrero, J., Abagyan, R., & Marsden, B. D. (2009). SGC—Structural biology and human health: A new approach to publishing structural biology results. *PLoS One, 4*(10), e7675.

Löser, P., Kobold, S., Guhr, A., Muller, F. J., & Kurtz, A. (2012). Scope and impact of international research in human pluripotent stem cells. *Stem Cell Reviews and Reports, 8*(4), 1048–1055.

Lowenthal, J., Lipnick, S., Rao, M., & Hull, S. C. (2012). Specimen collection for induced pluripotent stem cell research: Harmonizing the approach to informed consent. *Stem Cells Translational Medicine, 1*(5), 409–421.

Mathews, D. J. H., Cook-Deegan, R., & Bubela, T. (2013). Patents and misplaced angst: Lessons for translational stem cell research from genomics. *Cell Stem Cell, 12*(5), 508–512.
Mathews, D. J. H., Graff, G. D., Saha, K., & Winickoff, D. (2011). Access to stem cells and data: Persons, property rights, and scientific progress. *Science, 331*(6018), 725–727.
McCormick, J. B., Owen-Smith, J., & Scott, C. T. (2009). Distribution of human embryonic stem cell lines: Who, when, and where. *Cell Stem Cell, 4*(2), 107–110.
Miller, F. A., Sanders, C. B., & Lehoux, P. (2009). Imagining value, imagining users: Academic technology transfer for health innovation. *Social Science and Medicine, 68*(8), 1481–1488.
Mirowski, P. (2008). Livin' with the MTA. *Minerva, 46*(3), 317–342.
Moran, M., Guzman, J., Henderson, K., Abela-Oversteegen, L., Wu, L., Omune, B., et al. (2011). *Neglected disease research and development: Is the global financial crisis changing R&D?* Retrieved March 18, 2013, from http://www.policycures.org/downloads/g-finder_2010.pdf.
Moran, M., Guzman, J., Ropars, A.-L., McDonald, A., Jameson, N., Omune, B., et al. (2009). Neglected disease research and development: How much are we really spending? *PLoS Medicine, 6*(2), e1000030.
Mowery, D. C., & Ziedonis, A. A. (2007). Academic patents and materials transfer agreements: Substitutes or complements? *Journal of Technology Transfer, 32*(3), 157–172.
Munos, B. (2006). Can open-source R&D reinvigorate drug research? [10.1038/nrd2131]. *Nature Reviews Drug Discovery, 5*(9), 723–729.
Munos, B. (2009). Lessons from 60 years of pharmaceutical innovation. [10.1038/nrd2961]. *Nature Reviews Drug Discovery, 8*(12), 959–968.
Murray, F., Aghion, P., Dewatripont, M., Kolev, J., & Stern, S. (2009). *Of Mice and Academics: Examining the Effect of Openness on Innovation*. Working Paper No 14819, National Bureau of Economic Research. Retrieved March 11, 2013, from http://www.nber.org/papers/w14819.
Murray, F. (2010). The oncomouse that roared: Hybrid exchange strategies as a source of distinction at the boundary of overlapping institutions. *American Journal of Sociology, 116*(2), 341–388.
National Institutes of Health. (1995). *Master agreement regarding use of the uniform biological material transfer agreement*. Retrieved March 11, 2013, from http://www.ott.nih.gov/NewPages/UBMTA.pdf.
National Institutes of Health. (1999). Principles and guidelines for recipients of NIH research grants and contracts on obtaining and disseminating biomedical research resources. *Federal Register, 64*(246), 72090. Retrieved March 18, 2013, from http://grants.nih.gov/grants/intell-property_64FR72090.pdf.
Norman, T., Edwards, A., Bountra, C., & Friend, S. (2011). The precompetitive space: Time to move the yardsticks. *Science Translational Medicine, 3*(76), 76cm10.
Organisation for Economic Co-operation and Development. (2006). *Guidelines for the licensing of genetic inventions*. Retrieved March 11, 2013, from http://www.oecd.org/sti/biotech/36198812.pdf.
Organisation for Economic Co-operation and Development. (2009). *OECD patent statistics manual*. Retrieved March 11, 2013, from http://browse.oecdbookshop.org/oecd/pdfs/free/9209021e.pdf.
Organisation for Economic Co-operation and Development. (2011). *Collaborative mechanisms for intellectual property management in the life sciences*. Retrieved March 18, 2013, from http://www.oecd.org/sti/biotech/48665248.pdf.
Parsons, X. H., Teng, Y. D., Parsons, J. F., Snyder, E. Y., Smotrich, D. B., & Moore, D. A. (2011). Efficient derivation of human neuronal progenitors and neurons from pluripotent human embryonic stem cells with small molecule induction. *Journal of Visualized Experiments,* (56), 3273.
Paul, S. M., Mytelka, D. S., Dunwiddie, C. T., Persinger, C. C., Munos, B. H., Lindborg, S. R., et al. (2010). How to improve R&D productivity: The pharmaceutical industry's grand challenge. *Nature Reviews Drug Discovery, 9*(3), 203–214.
Pera, M. F. (2011). Stem cells: The dark side of induced pluripotency. *Nature, 471*(7336), 46–47.
Rai, A. K., & Eisenberg, R. S. (2004). Proprietary considerations. In R. Lanza, et al. (Eds.), *Handbook of stem cells* (Vol. 2, pp. 793–798). Retrieved March 14, 2013, from http://scholarship.law.duke.edu/cgi/viewcontent.cgi?article=2036&context=faculty_scholarship.

Rao, M. S. (2006). Embryonic stem cell research and U.S. Policy. *Stem Cells, 24*(6), 1412–1413.
Reed, W., Noga, S. J., Gee, A. P., Rooney, C. M., Wagner, J. E., McCullough, J., et al. (2009). Production assistance for cellular therapies (PACT): Four-year experience from the United States National Heart, Lung, and Blood Institute (NHLBI) contract research program in cell and tissue therapies. *Transfusion, 49*(4), 786–796.
Rodriguez, V. (2008). Governance of material transfer agreements. *Technology in Society, 30*(2), 122–128.
Rossant, J., & McMahon, A. (1999). 'Cre'-ating mouse mutants—A meeting review on conditional mouse genetics. *Genes and Development, 13*(2), 142–145.
Russell, S. (2010). Britain grants patent for iPS cells. *Nature News*. Retrieved March 11, 2013, from http://www.nature.com/news/2010/100128/full/news.2010.43.html.
Saha, K., Winickoff, D. E., & Graff, G. D. (2010). *Opening stem cell research and development: A policy proposal for the management of data, intellectual property, and ethics*. Retrieved March 11, 2013, from http://ourenvironment.berkeley.edu/wp-content/uploads/2011/09/Winickoff_YJHLPE.dpf_.pdf.
Schofield, P. N., Bubela, T., Weaver, T., Portilla, L., Brown, S. D., Hancock, J. M., et al. (2009). Post-publication sharing of data and tools. *Nature, 461*(7261), 171–173.
Schofield, P. N., Eppig, J., Huala, E., De Angelis, M. H., Harvey, M., Davidson, D., et al. (2010). Sustaining the data and bioresource commons. *Science, 330*(6004), 592–593.
Simon, B. M., Murdoch, C. E., & Scott, C. T. (2010). Pluripotent patents make prime time: An analysis of the emerging landscape. *Nature Biotechnology, 28*(6), 557–559.
Streitz, W. D., & Bennett, A. B. (2003). Material transfer agreements: A University perspective. *Plant Physiology, 133*(1), 10–13.
Trajtenberg, M. (1990). A penny for your quotes: Patent citations and the value of innovations. *The RAND Journal of Economics, 21*(1), 172–187.
UK Intellectual Property Office. (2010). *Overview of the UK national stem cell network patent watch landscape*. Retrieved March 18, 2013, from http://hinxtongroup.files.wordpress.com/2010/11/stemcellpatents_ukscn-2009-10_informatics_jul2010.pdf.
Voell, M. R., Farris, L., Levy, E., & Marden, E. (2010). A response to Rome: Lessons from pre- and post-publication data-sharing in the C. elegans research community. *BMC Genomics, 11*(1), 708.
von Tigerstrom, B. J. (2008). The challenges of regulating stem cell-based products. *Trends in Biotechnology, 26*(12), 653–658.
von Tigerstrom, B. (2011). The Food and Drug Administration, regenerative sciences, and the regulation of autologous stem cell therapies. *Food and Drug Law Journal, 66*, 479–506.
Vrtovec, K. T., & Vrtovec, B. (2013). Patent eligibility of induced pluripotent stem cell theranostics. *Clinical Pharmacology & Therapeutics, 93*(4), 318–320. doi:10.1038/clpt.2013.11.
Walsh, J. P., Cohen, W. M., & Cho, C. (2007). Where excludability matters: Material versus intellectual property in academic biomedical research. *Research Policy, 36*(8), 1184–1203.
Walsh, J. P., & Hong, W. (2003). Secrecy is increasing in step with competition. *Nature, 422*(6934), 801–802.
Weigelt, J. (2009). The case for open-access chemical biology. A strategy for pre-competitive medicinal chemistry to promote drug discovery. *EMBO Reports, 10*(9), 941–945.
Winickoff, D. E. (2006). Governing stem cell research in California and the USA: Towards a social infrastructure. *Trends in Biotechnology, 24*(9), 390–394.
Winickoff, D. E., Saha, K., & Graff, G. (2009). Opening stem cell research and development: A policy proposal for the management of data, intellectual property, and ethics. *Yale Journal of Health Policy, Law, and Ethics, 9*(1), 52–127.
Woodcock, J. (2010). Precompetitive research: A new prescription for drug development? [News]. *Clinical Pharmacology & Therapeutics, 87*(5), 521–523. doi:10.1038/clpt.2010.28.
Zarzeczny, A., Scott, C., Hyun, I., Bennett, J., Chandler, J., Chargé, S., et al. (2009). iPS cells: Mapping the policy issues. *Cell, 139*(6), 1032–1037.
Zwaka, T. P. (2010). Stem cells: Troublesome memories. *Nature, 467*(7313), 280–281.

Chapter 4
The Patenting Landscape for Human Embryonic Stem Cells

Kevin E. Noonan

4.1 Introduction

Patent protection in regenerative medicine, like patenting generally in biomedicine, is an area fraught with controversy over the past decade. For a time, patent eligibility, particularly in the USA, was broadly granted based on expansive language from the US Supreme Court's decision in *Diamond v. Chakrabarty*, 447 U.S. 303 (1980). Although the U.S. Patent and Trademark Office has relied on that language for more than 30 years, recently some have questioned the wisdom of providing patent protection for human-derived molecules. These include composition of matter claims to molecules such as DNA as well as claims to diagnostic methods. Opposition to such patents has arisen for religious and political grounds, but also based on a fear that patenting may inhibit innovation in this area. While there is little empirical support for this sentiment, the fear persists and can influence courts and policymakers (Heller and Eisenberg 1998).[1] In this chapter I provide a review of some of these issues and how they have affected patent protection in regenerative medicine. I begin by reviewing some key elements of patents, particularly as they relate to patenting living biological entities. A brief history of efforts to patent stem cells and reaction to these efforts follows, including the application of the US law from legislative actions regarding federal funding of stem cell research, Congressional patent reform activities and recent relevant court cases in the USA and Europe. Future prospects are also discussed.

[1] See also, SACGHS (2009); The Hinxton Group (2008a, b); Cook-Deegan et al. (2009).

K.E. Noonan (✉)
McDonnell Boehnen Hulbert & Berghoff LLP, Chicago Illinois
e-mail: noonan@mbhb.com

L.F. Hogle (ed.), *Regenerative Medicine Ethics: Governing Research and Knowledge Practices*, DOI 10.1007/978-1-4614-9062-3_4,

4.2 Patenting: A Brief Primer

4.2.1 The Basics of Patent Law

The basics of patent law have become increasingly harmonized between the USA and the rest of the world, particularly Europe, over the past several decades. Patents are granted on machines, manufactured goods, processes and compositions of matter that satisfy three basic patentability requirements: novelty, non-obviousness (in Europe, inventive step), and utility (or industrial applicability). In addition, all countries require disclosure of the invention in "such full, clear, concise, and exact terms as to enable any person skilled in the art to which it pertains, or with which it is most nearly connected, to make and use" it.[2] Until recently, the US law also contained a provision requiring disclosure of the "best mode" of making and using the invention (although this remains, failure to comply can no longer be the basis for invalidating a patent). This disclosure requirement is considered the "quid pro quo" of the patent grant, representing the exchange of full disclosure of the invention in return for the right to exclude others to make, use, sell, offer to sell or import a patented product or practice a patented method.[3]

The disclosure requirement is intended to provide two benefits to the public. The first is immediate: disclosure of the invention provides the skilled worker with information (which is not patented) that can be used either to "design around" a patented product or, more frequently, spark additional creativity and invention based on the advancement in the art the disclosed invention represents. The second benefit comes after the patent expires: because the inventor has satisfied the disclosure requirement (which is far stricter than the requirements for publication in a scientific journal, for example), the person skilled in the art to which the invention pertains is capable of practicing the invention throughout its full scope. The extent of this benefit varies with technology; it may not be as valuable for cell phones due to their relatively short obsolescence timelines, but for biotechnology, the pharmaceutical arts, and regenerative medicine, the technology generally holds its value for a much longer time.

This second benefit is also related to an often overlooked feature on patents (at least when any innovation-diminishing aspects are discussed): patents expire, generally 20 years after the earliest filing date of the first patent application that is filed on the technology.[4] Even that term requires payment of maintenance fees that increase over time. These fees are intended to provide an incentive for patentees to dedicate their patent rights to the public if the revenues generated from the technology a patent protects do not justify the cost of the maintenance fee payment. Thus, the term of a patent, while having a fixed expiration date, does not always survive to achieve that extent of the exclusionary period.

---

[2] 35 U.S.C. §112, first paragraph.

[3] 35 U.S.C. § 271(a).

[4] Because the 20 year term only came into effect in the U.S. after the adoption of the GATT/TRIPS agreement in 1995, there are some biotechnology patents having longer terms under the old regime of 17 years from grant date.

4.2.2 Patentable Subject Matter: "Anything Under the Sun Made by Man"

The modern era of patent protection for biotechnology began with the companion cases of *Diamond v. Chakrabarty* and *In re Bergy*.[5] While the *Chakrabarty* case was the more influential, setting a US Supreme Court precedent, Judge Giles Sutherland Rich's analysis in *Bergy* set forth important precedent relating to the patent eligibility of living things. Surveying more than 100 years of patents, from Louis Pasteur's yeast strain that saved the French wine industry in the Nineteenth Century to later patents on various microbial and other species, the *Bergy* opinion rejected protestations from the Patent Office that living things should not be patent-eligible per se and that provisions of patent law regarding plant protection (specifically, the Plant Protection Act of the Plant Variety Protection Act) indicated a Congressional intent *not* to patent other life forms.

The Supreme Court came to a similar conclusion: while *Bergy* involved the patent eligibility of a "biologically pure culture" of a naturally occurring bacterium (*Streptomyces vellosus*, that produced the antibiotic lincomycin), the *Chakrabarty* invention was a genetically engineered *Pseudomonas aeruginosa* bacteria capable of metabolizing components of crude oil, intended to provide a biological ability to clean up oils spills in the ocean.[6] The Court's decision, that Chakrabarty's bacteria was patent-eligible, was based on the perception that the difference between the patent-eligible and non-eligible depended on whether there was significant evidence of the "hand of man" that transformed the bacteria into something that did not exist in nature. The Court held that Chakrabarty's bacteria was patent-eligible because the subject matter could be distinguished from earlier precedent barring patenting of laws of nature or that did not show sufficient evidence of the "hand of man" to pass patent eligibility muster.[7]

However, it was the expansive scope of the *Chakrabarty* decision that was most significant. The Court found that Congress had evinced a desire that patent eligibility should be construed broadly, because the statute used the word "any" to modify the particularly enunciated scope of patent-eligible subject matter, and the legislative history was in accord, stating that Congress's intent was that "anything under the sun made by man" would be patent-eligible. Although the decision was only 5-4, the effect of the Court's language was to encourage the Patent Office to remove

[5] 596 F.2d 952 (C.C.P.A. 1979).

[6] The bacterium was not a recombinant organism, i.e., no novel recombinant genetic construct was introduced. Rather, using conventional transduction protocols *Chakrabarty* introduced three naturally occurring plasmids into the bacteria and selected a stable strain. These plasmids encoded members of a metabolic pathway that contributed to digestion of the crude oil components by the bacteria.

[7] In the former case, *Funk Bros. Seed Co. v. Kalo Inoculant Co.*, 333 U.S. 127 (1948), the claims recited any combination of *Rhizobium* bacteria that could be cultured together; claims in the latter case, *Am. Fruit Growers v. Brogdex Co.*, 283 U.S. 1 (1931), were directed to borax-impregnated oranges.

most restrictions on subject matter eligibility. Coupled with the *Bergy* decision (in which the PTO's supervisory court deemed the "biologically pure culture" of a naturally occurring bacterium *not* to be a "product of nature"), the Office's practice was to permit patents on "isolated" or "isolated and purified" genes, proteins, cells, metabolites, and other products of biological processes. This included multicellular organisms, most significantly a mouse that expresses an oncogene (*c-myc*) that confers on the mouse susceptibility for breast cancer (commonly termed the "Harvard oncomouse," U.S. Patent No. 4,736,866).[8] These claims had in common that they were required to recite that the claimed subject matter was "isolated" (or alternatively, "purified" or "isolated *and* purified"), and for claims to animals to recite a "non-human animal," in deference to the provisions of the 13th Amendment. This PTO action expanded the scope of patentable organisms from the microscopic to the mammalian, generating significant controversy, particularly in foreign countries.[9]

Legal challenges to these practices were few and generally dismissed on procedural grounds. During this period, several thousands of patents on the products of the biotechnology revolution were routinely granted and, generally, unchallenged in the courts. However, many commentators questioned the patent eligibility of genes and other biological molecules, often based on perceived threats to innovation that these patents could present (Heller and Eisenberg 1998).

4.3 Patenting Stem Cells

Patented stem cells initially fit easily into the statutory, administrative, and expansive judicial framework engendered by the *Chakrabarty* decision. The first stem cell patents were directed to hematopoietic stem cells (e.g., U.S. Patent No. 5,436,151 and 5,670,147), fetal/neonatal cells (e.g., U.S. Patent No. 5,004,681), and mesenchymal cells (e.g., U.S. Patent No. 5,827,740). Embryonic stem cells were first patented from animals (e.g., birds (U.S. Patent Nos. 5,340,740 and 5,656,479) and mice (U.S. Patent Nos. 5,453,357 and 5,985,659)). The first human embryonic stem cell patents were granted to James Thomson from the University of Wisconsin (U.S. Patent Nos. 5,843,780; 6,200,806; 7,029,913; *see*, Table 7.1), and in many respects, represent the "high water mark" of claiming biotechnological inventions, being directed not just to biological material but to cells that, under the right conditions, could be used to create a human being. Since the first of these patents, more than 1,000 patents claiming stem cells have been granted in the USA. Worldwide, there are several thousand patents on human and nonhuman stem cells, as well as a smaller number of patents on embryonic stem cells (Table 4.1).

---

[8] Claim 1. A transgenic non-human mammal all of whose germ cells and somatic cells contain a recombinant activated oncogene sequence introduced into said mammal, or an ancestor of said mammal, at an embryonic stage.

[9] For example, the Canadian Supreme Court overruled the Canadian Patent Office and lower courts and required cancelation of claims to the mouse itself, while in Europe the patent was ultimately revoked after extensive opposition proceedings by 17 separate opponents.

Table 4.1 First embryonic stem cell patents (J. Thomson)

U.S. Patent No. 5,843,780	U.S. Patent No. 6,200,806	U.S. Patent No. 7,029,913
A purified preparation of primate embryonic stem cells which (i) is capable of proliferation in an in vitro culture for over 1 year, (ii) maintains a karyotype in which all the chromosomes characteristic of the primate species are present and not noticeably altered through prolonged culture, (iii) maintains the potential to differentiate into derivatives of endoderm, mesoderm, and ectoderm tissues throughout the culture, and (iv) will not differentiate when cultured on a fibroblast feeder layer	A purified preparation of pluripotent human embryonic stem cells which (i) will proliferate in an in vitro culture for over 1 year, (ii) maintains a karyotype in which the chromosomes are euploid and not altered through prolonged culture, (iii) maintains the potential to differentiate to derivatives of endoderm, mesoderm, and ectoderm tissues throughout the culture, and (iv) is inhibited from differentiation when cultured on a fibroblast feeder layer	A replicating in vitro cell culture of human embryonic stem cells comprising cells which (i) are capable of proliferation in in vitro culture for over 1 year without the application of exogenous leukemia inhibitory factor, (ii) maintain a karyotype in which the chromosomes are euploid through prolonged culture, (iii) maintain the potential to differentiate to derivatives of endoderm, mesoderm, and ectoderm tissues throughout the culture, and (iv) are inhibited from differentiation when cultured on a fibroblast feeder layer

4.4 Let the Backlash Begin

The reaction to such patents, and human embryonic stem cell technology generally, was not altogether positive. In a national address on August 9, 2001, President George W. Bush established a ban on Federal government funding for stem cell research with the exception of a set of identified, preexisting stem cell lines. Permitting the use of these lines in principle allowed some research to proceed, while preventing the creation and use of new lines, and was meant to satisfy those who opposed the use of embryos for research. Yet most of the designated lines could not be used for research, either because they were not made available to scientists, were not true cell lines, or otherwise were not usable (Murugan, 2009).[10] As a consequence, many advances in stem cell research during this period were made abroad, in the United Kingdom, Singapore, Australia, and other countries that permit embryos to be used for research. Several states (including California and Massachusetts) established mechanisms for funding embryonic stem cell research, with the effect of differing policies for funding and executing research within a single country. As described in the next section, the Obama Administration changed federal funding policy and guidelines for research.[11] Stem cell research has been supported in some countries and opposed in others (See Appendix A). In Europe patenting stem cells have been opposed several times.[12]

[10] For discussions of the politics of the decision, see, for example, Shulman (2006), Mooney (2006) and Kass (2002).

[11] Executive Order 13505, "Removing Barriers to Responsible Scientific Research Involving Human Stem Cells" (March 9, 2009).

[12] *See infra* at "7.5.3 The Rest of the World: What's the Situation in Europe?"

The most significant challenge to stem cell patents in the USA was raised by the Public Patent Foundation (PubPat), a public interest group headed by Daniel Ravicher, a law professor at Cardozo School of Law in New York, joined by The Foundation for Taxpayer and Consumer Rights (FTCR) (now known as Consumer Watchdog), a California taxpayer group. Both *ex parte* and *inter partes* reexamination requests were filed (and granted) against the Thomson patents.[13] The Office initially rejected the claims of each of the patents in reexamination; this was not surprising, since it is rare for a reexamination request to be granted unless there are objective grounds for a rejection.[14] The patent holder, the Wisconsin Alumni Research Foundation (WARF), responded by amending claim 1 of the '608 patent (directed to the stem cells themselves), to recite that the stem cells were "derived from a preimplantation embryo." WARF also amended claim 9 (directed to methods for producing the stem cells) to recite that the cells produced by the method were "capable of proliferation as undifferentiated cells for more than 1 year." The '806 patent's claims were similarly amended. WARF's arguments were buttressed by an expert declaration from a mouse embryologist, who averred to failure of others to isolate human and nonhuman stem cells. The Patent Office determined that the amended claims of both patents remained patentable after reexamination,[15] and in due course issued reexamination certificates for both patents.

A similar outcome was achieved for the *inter partes* reexamination at least initially.[16] Rejecting the grounds of unpatentability asserted by the requestors, the Office asserted its own grounds of rejection based on lack of novelty and obviousness.[17] WARF responded as it had in the *ex parte* reexaminations. However, the requestors had the opportunity to comment on WARF's responses, by submitting competing expert declarations of their own. These declarations, made by stem cell researchers, asserted that Dr. Thomson was able to produce human embryonic stem cells because he had unique access to human embryos obtained from an Israeli scientist, as well as funding from Geron, a California company that had exclusively

[13] The *ex parte* reexaminations against U.S. Patent No. 5,843,780 and 6,200,806 were assigned Control Nos. 90/008102 and 90/008139, respectively, and the *inter partes* reexamination against U.S. Patent No. 7,029,913 was assigned Control No. 95/000154. The complete file histories of these reexaminations can be viewed under the Public portion of the Patent Application Information Retrieval (PAIR) system at www.uspto.gov.

[14] The grounds of rejection asserted by the Patent Office in reexamination of "the '780 patent that the claims were unpatentable for lack of novelty under 35 U.S.C. § 102(b) or for being obvious under 35 U.S.C. § 103(a) over a variety of references that were published prior to the priority date of the Thomson patents.

[15] *Supra* note 13.

[16] The grounds asserted by the requestors in the *inter partes* reexamination were slightly different than in the *ex parte* reexaminations, due to the requestors' ability to submit affidavit evidence *inter partes*. The grounds of unpatentability asserted were supported by declaration evidence and the prior art references cited in the parallel *ex parte* reexaminations. *See supra* note 13, *supra*.

[17] The Office expressly rejected the assertions in the declaration as the basis for its determination, stating that it expressed "opinion on the ultimate legal issue without providing sufficient underlying factual support." *Supra* note 13.

licensed WARF's stem cell patents in certain fields of application. They argued that Dr. Thomson succeeded in isolating human ES cells where others failed merely due to access to resources (human embryos) and funding others did not have. The Examiner determined that this was merely hindsight reasoning and thus impermissible for consideration in a determination of obviousness. The Office withdrew the asserted grounds of rejection and refuted or rebutted the comments made by the requestors in finding the claims allowable. In doing so, the Office cited numerous prior art references supporting the non-obviousness of the Thomson claims. This evidence supported the Examiner's position that the art showed that isolating human ES cells were sufficiently unpredictable to amount to undue experimentation (unless and until ES cells from a particular species had been made).

However, the requestors appealed to the Board of Patent Appeals and Interferences, who rendered a decision overturning the Examiner's patentability decision on the obviousness issue, based on the Board's determination that it would have been "obvious to try" to obtain human embryonic stem cells in view of the successes with other mammalian species.[18] The Board also was not convinced that production of human embryonic stem cells was not sufficiently unpredictable to render the claims nonobvious. On remand, the Examiner duly asserted these grounds, and WARF responded with evidence and argument rebutting the grounds of rejection. In November, 2011, the Examiner once again determined that the claims were patentable over the prior art of record but has not entered an Action closing prosecution (necessary to trigger the requestors' right to file another appeal).

The ultimate fate of the WARF/Thomson patents may not be determined before they expire in 2015. Moreover, the reexaminations were limited to reconsideration based, as is statutorily required, on prior patents and publications and allegations of obviousness and lack of novelty. Since these reexaminations were filed, additional grounds for attacking stem cell patents and patents claiming other human-derived biological molecules (specifically, isolated DNA) have been mounted by PubPat and the American Civil Liberties Union based on these molecules not being statutory subject matter. One lawsuit in particular, *Association of Molecular Pathologists et al. v. U.S. Patent and Trademark Office* et al. (involving Myriad Genetics), has raised the possibility that the broad scope of patent eligibility created by the *Chakrabarty* decision is open to reconsideration by the current Supreme Court and indeed the Supreme Court has limited the scope of subject matter eligibility in its decision in this case.[19] Other developments, in Congress and the Obama Administration, also suggest that patent scope is being seriously reconsidered. Key cases and developments will be discussed in the next section.

[18] In its decision, the Board reinstated rejection for obviousness under 35 U.S.C. §103(a) on the following grounds: (1) that claims 1-3 were unpatentable over U.S. Patent No. 5,166,065 to Williams; (2) that claims 1-3 were unpatentable over the combination of the Robertson '83 reference, the Robertson '87 reference, the '065 patent, and the '926 patent; (3) that claims 1-3 were unpatentable over the combination of the Piedrahita reference, the '065 patent and the '926 patent; and (4) that claims 1-3 were unpatentable over the combination of the Robertson '83 reference, the Robertson '87 reference, the Piedrahita reference, the '065 patent, and the '926 patent.

[19] 702 F. Supp.2d 181 (S.D.N.Y. 2010).

4.5 What's Happening Now

Each of the three branches of government in the USA, the executive, legislative, and judicial, are in a position to influence stem cell policy and the extent to which the law promotes or inhibits stem cell research. While there have been no changes specific to stem cell patents, there are a few troubling trends that may raise issues regarding the continued scope or even eligibility of stem cell patents.

4.5.1 The Executive Branch

Guidelines promulgated by the National Institutes of Health regarding Federally funded stem cell research have expanded the number of approved human embryonic stem cell lines from 69 to well over 1,000 and have made a more favorable climate for deriving new lines, greatly increasing the potential for more rapid development of the technology (and concomitant creation of patents and other IP rights).[20] The Guidelines were promulgated pursuant to Executive Order 13505 on March 9, 2009, entitled "Removing Barriers to Responsible Scientific Research Involving Human Stem Cells." These Guidelines establish a stem cell registry administered by the National Institutes of Health and provide certain restrictions for cells entered into the registry (but many fewer than under the Bush administration).[21] The Guidelines continue to prohibit Federal funding for research that in any way introduces human embryonic stem cells into a nonhuman animal for breeding experiments. There are no provisions for "grandfathering" already established cell line: these are being individually evaluated under ethical principles contained in the Guidelines as well as 45 CFR Part 46 Subpart A (Protection of Human Subjects). The Guidelines also assert that the Institute expects "that stem cell research materials developed with NIH funds, as well as associated intellectual property and data, will be distributed in accordance with the NIH's existing policies and guidance, including 'Sharing Biomedical Research Resources, Principles and Guidelines for Recipients of NIH Grants and Contracts' and 'Best Practices for the Licensing of Genomic Inventions'" (see NIH Policies & Reports page). While these developments are encouraging, they may be ephemeral as political and social climates change with each election cycle. New administrations have substantial power to oppose or restrict stem cell research.

The U.S. Patent and Trademark Office (a unit of the Department of Commerce) has not changed its policies concerning patents claiming human stem cells or their uses, including embryonic, adult and tumor-derived, human and nonhuman cells.

---

[20] National Institutes of Health Guidelines for Human Stem Cell Research, 74 Fed. Reg. 32,170 (July 7, 2009).

[21] The Guidelines also provide a shortened path for cells established before the Guidelines were promulgated, requiring that the cells were created using in vitro fertilization for reproductive purposes and were no longer needed for this purpose; and were donated by donor(s) who gave voluntary written consent for the human embryos to be used for research purposes.

In fact, some two dozen patents were granted in 2011–2012.[22] However, the U.S. Department of Justice, in its amicus brief in the *Myriad* case, has taken the position that "products of nature" that are merely isolated and not otherwise changed should be ineligible for patenting. This posture would seem to preclude patent protection for human stem cells.

4.5.2 *The Legislative Branch*

After a decade of failed attempts, Congress passed comprehensive patent reform in the USA in 2011.[23] There are no provisions that expressly recite stem cells (human or otherwise), but Section 30 of the law as enacted includes a provision (to be codified as part of 35 U.S.C. §101) that may impact the patent eligibility of human stem cell claims:

Sec. 30. LIMITATION ON ISSUANCE OF PATENTS

(a) LIMITATION: Notwithstanding any other provision of law, no patent may issue on a claim directed to or encompassing a human organism.
(b) EFFECTIVE DATE:
    (1) IN GENERAL: Subsection (a) shall apply to any application for patent that is pending on, or filed on or after, the date of enactment of this Act
    (2) PRIOR APPLICATIONS: Subsection (a) shall not affect the validity of any patent issued on an application to which paragraph (1) does not apply.

This change in the law codifies the Weldon amendment, an amendment added to every appropriations bill since 2004 which states: "None of the funds appropriated or otherwise made available under this Act may be used to issue patents on claims directed to or encompassing a human organism."[24] That amendment had been routinely accompanied by assurances that it was not intended to limit patenting of genes or other products of biotechnology, as evidenced, *inter alia*, by a November 20, 2003 letter from Senator Ted Stevens (R-AK) to USPTO Director Rogan and colloquy on the House floor between Reps. Weldon (15th Dist. FL) and

---

[22] Patents granted in 2011 and 2012 include U.S. Patent Nos. 7,932,084 and 8,119,398 issued on 2/21/2012 (adipose stem cells), 7,947,501 (recombination in human ES cells), 7,955,846 (female germline stem cells), 7,955,850 (dental pulp-derived stem cells), 7,964,402 (clonal populations of hESCs), 7,968,336 (fetal stem cells), 7,968,337 (neural stem cells), 7,972,767, 8,084,023, and 8,142,773 issued on 3/27/2012 (mesenchymal stem cells), 7,976,836 and 8,057,789 (placental stem cells), 7,994,144 and 8,017,393 (hematopoietic stem cells), 8,034,329 (umbilical cord stem cells), 8,110,399 issued on 2/7/2012 (islet of Langerhans stem cells), 8,153,359 issued on 4/10/2012 (toxicity studies in human ESCs) and 8,153,421 issued on 4/10/2012 (prostate stem cells).

[23] The Leahy Smith America Invents Act, Pub. L. No. 112-29. (September 16, 2011).

[24] *See, for example*, Consolidated Appropriations Act of (2010), Pub. L. NO. 111-117, 123 Stat. 3034, 3153.

Obey (7th Dist. WI).[25] Specifically excluded from the scope of this exemption have been the following:

1. Any chemical compound or composition, whether obtained from animals or human beings or produced synthetically, and whether identical to or distinct from a chemical structure as found in an animal or human being, including but not limited to nucleic acids, polypeptides, proteins, antibodies, and hormones;
2. Cells, tissue, organs, or other bodily components produced through human intervention, whether obtained from animals, human beings, or other sources; including but not limited to stem cells, stem cell-derived tissues, stem cell lines, and viable synthetic organs;
3. Methods for creating, modifying, or treating human organisms, including but not limited to methods for creating embryos through in vitro fertilization, methods of somatic cell nuclear transfer, medical or genetic therapies, methods for enhancing fertility, and methods for implanting embryos;
4. A nonhuman organism incorporating one or more genes taken from a human organism, including but not limited to a transgenic plant or animal, or animal models used for scientific research.

However, it is unclear whether these Congressional caveats will withstand the expected court challenge that human-derived biomolecules should fall within the scope of this exclusion. For now, it may be enough that similar limitations on patenting human beings are already part of the U.S. Patent Office practice[26] and thus that this portion of the statute should be interpreted merely to maintain the status quo.[27]

4.5.3 *The Judicial Branch*

Like the other branches, the issue of stem cell patent eligibility has not arisen in US courts except for a failed attempt to prevent Federal funding of stem cell research *Sherley v. Sebelius*, 644 F.3d 388 (D.C. Cir. 2011). But two life sciences cases threaten a shift in the patenting zeitgeist that could raise additional impediments to human stem cell patenting in the USA.

The first of these, *Mayo Collaborative Services v. Prometheus Labs., Inc.*, was decided in 2012.[28] The case involved a method for optimizing therapeutic drug administration by testing patient blood for the presence of the drug or its metabolites. Specifically, the drug thiopurine would be given, metabolites from the drug made by the body were then measured, and the information used to adjust the

[25] Speech of Hon Lamar Smith of Texas, 112 *Cong. Rec.* E1182-E1185 (June 22, 2011).

[26] See Manual of Patent Examining Procedure 8th Ed. § 2105 (Revised August 2012).

[27] "If the broadest reasonable interpretation of the claimed invention as a whole encompasses a human being, then a rejection under 35 U.S.C. 101 must be made indicating that the claimed invention is directed to nonstatutory subject matter." *Id*

[28] 132 S.Ct. 1289 (2012).

administered dose if it fell outside a specific therapeutic window. The claims were found to be ineligible for patenting because the Court emphasized that "laws of nature" (and by implication, for some, "products of nature") are not eligible for patenting per se. The Court further seemed concerned with allegations by medical groups and others that such patents retarded innovation and prevented doctors from providing medical care to patients. Justice Breyer, writing for a unanimous Court, based his conclusion that the Prometheus claims were not patent-eligible by reasoning that the claims recited nothing more than a law of nature coupled with "routine, conventional and well-known" methods. According to the Court's thinking, permitting claims to be patentable risked tying up basic tools of scientific and technological work so that the public would not be able to use these tools for further and future innovation. These sentiments would appear to be ripe for exploitation by groups who oppose human stem cell patenting in that it would provide a rationale for the Court to find against stem cell patents on similar grounds.

The other case, *Association of Molecular Pathologists v. U.S. Patent and Trademark Office*, involved claims to isolated DNA molecules encoding human BRCA genes and method claims for determining risk for developing breast and ovarian cancer based on the presence of particular mutations in the BRCA genes.[29] The district court ruled that both the isolated DNA claims and the diagnostic method claims were not eligible for patenting. The Federal Circuit reversed the decision as to the isolated DNA claims but affirmed as to the method claims. The plaintiffs petitioned for Supreme Court review and the Court granted their petition, vacated the Federal Circuit opinion, and remanded the case to the appellate court for reconsideration in view of the *Mayo* decision. While the Federal Circuit's earlier decision was firmly grounded in the prevailing Supreme Court precedent (including *Chakrabarty*), the case was sent back to the lower court for reconsideration in view of the Court's *Mayo* decision is troubling. Plaintiffs in the case, in earlier argument and particularly in more recent briefing and oral argument before the Federal Circuit, have argued that isolated DNA is not sufficiently distinctive from native DNA to be patent-eligible; these arguments were based primarily on the identity between the sequence of the native and isolated DNA. Remand also suggested that the Court may be considering a further expansion of its recent predilection to reduce patent rights.[30] The Federal Circuit has once again ruled, substantially along the same lines of reasoning used in its initial opinion, that isolated DNA comprises

[29] 669 F.Supp. 2d 365 (S.D.N.Y. 2009), affirmed in part, reversed in part, 653 F.3d 1329 (Fed. Cir. 2011), certiorari granted, decision vacated and remanded, 2012 U.S. LEXIS 2356 (U.S. Mar. 26, 2012).

[30] See, for example, the Court's decisions in *KSR Int'l Co. v. Teleflex Inc* 127 S.Ct. 1727 (2007). (expanding obviousness), *eBay Inc. v. MercExchange LLC* 547 U.S. 388 (2006) (reducing injunctions), *Quanta Computer, Inc. v. LG Electronics, Inc* 553 U.S. 617 (2008). (expanding scope of patent exhaustion), *Merck KGAA v. Integra Lifesciences I, Ltd.* 545 U.S. 193 (2005) (expanding scope of "safe harbor" for FDA submissions), *Microsoft Corp. v. AT&T Corp.* 550 U.S. 437 (2007) (reducing scope of extraterritorial reach of U.S. patent law), *Festo Corp. v. Shoketsu Kinzoku Kogyo Kabushiki Co.*, 535 U.S. 722 (2002) (reducing scope of doctrine of equivalents, i.e., the extent to which insignificant changes to a claimed invention will still invoke patent infringement liability), and *Medimmune, Inc. v. Genentech, Inc.* 549 U.S. 118 (2007) (expanding declaratory judgment jurisdiction for challenging patents).

statutory subject matter under 35 U.S.C. §101. *Association for Molecular Pathologists* et al. *v. U.S. Patent and Trademark Office* et al., 689 F.3d 1303 (Fed. Cir. 2012).

One significant voice before the Federal Circuit was the US government by way of an amicus curiae brief to the court by the Department of Justice.[31] The Justice Department recommended that the court find that certain forms of DNA, specifically complementary DNA (cDNA) that is enzymatically synthesized from cellular messenger RNA (mRNA, should be patent-eligible, but genomic DNA and even synthetic oligonucleotides not be patent-eligible as being "merely" isolated products of nature. The government took the position that if a DNA molecule could be observed in nature (using a hypothetical "magic microscope") then that DNA should not be patented.

The Supreme Court granted *certiorari* on the question of "Are human genes patentable," and at the end of the recent term ruled substantially as advocated by the government.[32] According to the Court, cDNA satisfied patent eligibility requirements while genomic DNA did not. The Court used as its rationale that permitting genomic DNA to be patented would inhibit innovation by precluding research on such DNA, contravening its earlier precedent in *Funk Bros. Seed Co. v. Kalo Inoculant Co.*, 333 U.S. 127 (1948).

Thus, while there have been few policy decisions, laws, or court decisions expressly directed to human stem cells, it is clear that this recent decision by the US Supreme Court could be applied to stem cells (as well as all other biologically derived inventions) to be much more restrictive of exclusive rights for human-derived materials.[33] This represents a dramatic shift from the rationale from the *Chakrabarty* case and the expansive scope of protection that arose as a result of that decision.

4.6 The Rest of the World: What's the Situation in Europe?

The member states of the European Union are anything but united on the question of human embryonic stem cell research and patenting, with several countries banning the practice (Austria, France, Germany, Italy, The Netherlands, and Spain) while others do not (Belgium, Sweden, and the United Kingdom).

Patent protection for human stem cells in Europe recently became much more uniform, however, with a decision from the European Court of Justice (ECJ) that a procedure involving removal of stem cells from a blastocyst-stage embryo cannot be patented.[34] That decision under the "public morality" provisions of the EU

[31] The Patent and Trademark Office did not join in this brief, and it is widely believed that the Office does not agree with the brief's recommendations or conclusions.

[32] *Ass'n for Molecular Pathology v. Myriad Genetics, Inc.*, 133 S. Ct. 2107 (2013).

[33] Whether alternative embodiments, such as pluripotent stem cells produced by specific treatments of ESCs, would fall within any such preclusive determination of patent eligibility is unclear.

[34] Oliver Brüstle v. Greenpeace eV, European Court of Justice C-34/10, (Oct. 18, 2011).

heeded the recommendation of the court's advocate general that hESCs are not patent-eligible subject matter.[35] The case began in 2004, when Greenpeace sued in German federal court over a German patent to the University of Bonn involving methods for deriving neural cells from hESCs (DE 197568664 C1). While German laws regarding stem cell research have been characterized as the most restrictive in Europe, research is permitted provided that it is performed with pluripotent (rather than totipotent) cells, using cell lines not derived in Germany, and only cell lines that were made prior to May 2007.[36] Nevertheless, Greenpeace argued that claims to methods for using hESCs were "immoral and against public order." There is no corresponding provision under the US law.

The German federal court ruled for Greenpeace in 2006, and Dr. Brüstle and the University of Bonn appealed to the German Supreme Court. That court referred the question to the ECJ, since German law was closely patterned on European Union guidelines for biotechnology patenting. In the "Questions referred" to the ECJ, the German court asked for clarification on the meaning of the term "human embryos" in Article 6(2)(c) of Directive 98/44/EC of the European Parliament and in the Council of 6 July 1998 statement on the legal protection of biotechnological inventions (OJ 1998 L 213, p. 13). Clarification was also requested regarding the expression "uses of human embryos for industrial or commercial purposes," specifically whether this includes any commercial exploitation within the meaning of Article 6(1) of the Directive, especially use for the purposes of scientific research. Finally, the court asked whether methods for producing hESCs or using them for technical purposes are unpatentable. Specifically, the court requested guidance for situations where the use of human embryos was not part of the technical teaching claimed with the patent, but use of human embryos is a necessary precondition for the application of that teaching. These situations could arise, for example, "because the patent concerns a product whose production necessitates the prior destruction of human embryos or [] because the patent concerns a process for which such a product is needed as base material."

On March 10, 2011, the ECJ advocate-general, Judge Yves Bot, rendered an opinion that human embryonic stem cell patents were "contrary to ethics and public policy" because they required "industrial use" of human embryos. This recommendation was not binding on the ECJ, but it was expected that the court would agree with the advocate-general, since it is rare that such preliminary opinions are overruled. In its decision, the ECJ cited the Directive on Biotechnology as well as the relevant provisions of Articles 52(1) and 53 of the Convention on the Grant of European Patents relating to broad scope of patent protection (limited, *inter alia*, by consideration of "ordre public" or morality). The court also considered German law (Paragraph 2 of the Patentgesetz) as amended to comply with the Directive, and specifically the prohibition under German law of patenting "uses of human embryos

[35] Opinion of the Advocate General, Case C-34/10, Brüstle v. Greenpeace, (March 10, 2011).

[36] Germany, *Bundestag, Stem Cell Act of 2002* (Stammzellgesetz), Bundesgesetzblatt [Federal Law Gazette] 2002, Part I, no. 42, p. 2277, (June 29, 2002), §1-1.

for industrial or commercial purposes," and provisions of German law restricting the uses to which embryos can be put outside the patent context (such as laws against human cloning). Finally, the Court considered Europe's international responsibilities under the Trade-Related Aspects of Intellectual Property Rights (TRIPS) agreement (an international agreement administered by the World Trade Organization), and specifically the provisions of Article thereof that permits member states to limit the scope of patent protection in some circumstances.[37] In its decision, the ECJ held that "any human ovum must, as soon as fertilized, be regarded as a 'human embryo' within the meaning and for the purposes of the application of Article 6(2)(c) of the Directive, if that fertilization is such as to commence the process of development of a human being."[38] In addition, the court put into the "human embryo" category cloned ova (wherein the nucleus of a somatic cell is introduced into an enucleated ovum) and ova stimulated to divide parthenogenetically. The court then decided that the capacity for a totipotent or pluripotent human embryonic stem cell to develop into many or all human tissues was sufficient for them to qualify as a "human embryo".

The court further found that uses of hESCs for scientific research are encompassed (and precluded) by the Directive insofar as the use is subject to patent protection, because "clearly the grant of a patent implies, in principle, its industrial or commercial application." *Brüstle v. Greenpeace*, Case C-34/10, Oct. 18, 2011, at para. 41. In making this determination, the ECJ distinguished scientific research usage from uses for therapeutic or diagnostic purposes directed to the embryo, which are entitled to patent protection under other provisions of the Directive. The court also answered the question of whether a process is patent-ineligible where the "purpose is not the use of human embryos, where it concerns a product whose production necessitates the prior destruction of human embryos or a process for which requires a base material obtained by destruction of human embryos," (*Id.* at para. 47) holding that such processes are patent-ineligible because "the removal of a stem cell from a human embryo at the blastocyst stage entails the destruction of that embryo": (*Id.* at para. 48).

Accordingly, on the same grounds as those set out in paragraphs 32 to 35 above, an invention must be regarded as unpatentable, even if the claims of the patent do not concern the use of human embryos, where the implementation of the invention requires the destruction of human embryos. In that case too, the view must be

[37] Article 27 reads as follows: "Members may exclude from patentability, inventions, the prevention within their territory of the commercial exploitation of which is necessary to protect *ordre public* or morality, including to protect human, animal or plant life or health or to avoid serious prejudice to the environment, provided that such exclusion is not made merely because the exploitation is prohibited by law." Agreement on Trade-related Aspects of Intellectual Property Rights, Part II, Section 5, Article 27. Individual countries can make their own decisions based on moral objections (or acceptance) of a technology; for example, Sweden granted a patent for human ES cells (SES536490) based on the argument that while producing human ES cells was required, the cells may have been isolated long before the invention was made (and hence embryos were not instrumentalized or did not require repeated use of embryos (an element in Swedish patent law). See Torremans (2010. p. 299) and Salter (2007).

[38] Oliver Brüstle v. Greenpeace eV, European Court of Justice C-34/10, para. 35 (Oct. 18, 2011).

taken that there is use of human embryos within the meaning of Article 6(2)(c) of the Directive. The fact that destruction may occur at a stage long before the implementation of the invention, as in the case of the production of embryonic stem cells from a lineage of stem cells the mere production of which implied the destruction of human embryos is, in that regard, irrelevant." *Oliver Brüstle v. Greenpeace eV*, European Court of Justice C-34/10, para. 49 (Oct. 18, 2011).

The ECJ's decision is binding on the member states of the European Community. However, more recently on 27 November 2012, the German court determined that in vitro cells derived from the blastocyst stage of embryo development did not themselves have the capability to develop into people, and therefore did not count as human embryos, thereby distinguishing the ECJ's ruling. It remains to be seen whether this is the beginning of a trend in national courts, and whether Greenpeace and the other plaintiffs will be able to successfully challenge such rulings either nationally or before the ECJ.

4.7 Prospects for the Future

The ECJ's decision banning patenting of human stem cells, or any application of stem cell technology in Europe, is the latest in a series of events that may strengthen initiatives by individuals and groups with philosophical objections to patents involving biological materials, particularly material derived from human beings. These include the aforementioned Public Patent (PUBPAT) Foundation's attempts to invalidate the Thomson hESC patents in the USA, as well as PUBPAT and the ACLU's challenge to human gene patenting in the *AMP v. USPTO* (Myriad) case. Opponents of such patents voice various arguments and rationales against patenting products of human biology, including the claim that patenting inhibits basic research. Their opposition is directed towards patent-ineligible subject matter or has been aggressively licensed and enforced, resulting in restricted access based on price. Ultimately, however, all arguments against human stem cell patenting have to do with the fact that there are moral objections to the practice, as evidenced by the ECJ's decision. Such objections are sincerely held by many and it is neither possible nor productive to make arguments on these grounds. As it is in many public policy debates, these positions tend to be absolutes for some, not subject to merely reasoned argument.

Perhaps the best rejoinder of arguments opposing patenting of stem cell products is to posit the situation if patent opponents prevail. The resulting lack of patent protection can be expected to have two concrete and predictable consequences: lack of financial investment to translate basic scientific discoveries to useful commercial products, and (where possible) reducing or eliminating public disclosure of inventions (and where not possible, investment in other technologies). Neither outcome is conducive to reducing human morbidity or mortality or improving the human condition. It is hard to understand how advocating such an outcome can be considered the more moral position.

References

112 Cong. Rec. E1182 (June 22, 2011) (speech of Hon. Smith).

Bush, G. W. (2001). Presidential Address. August 9, 2001.

California Institute for Regenerative Medicine. (2009). *CIRM stem cell grant regulations*. Retrieved August 10, 2012, from http://www.cirm.ca.gov/cirm-operations/Regulations.

Cook-Deegan et al. (2009). The dangers of diagnostic monopolies. *Nature, 458*(7237), 405–406.

Executive Order 13505, "Removing Barriers to Responsible Scientific Research Involving Human Stem Cells" (March 9, 2009).

Heller, M. A., & Eisenberg, R. S. (1998). Can patents deter innovation? The anticommons in biomedical research. *Science, 280*, 698–701.

The Hinxton Group. (2008a). *The science, ethics, and policy challenges of pluripotent stem cell-derived gametes*. Retrieved August 8, 2012, from http://www.hinxtongroup.org/au_pscdg_cs.html.

The Hinxton Group. (2008b). *Transnational cooperation in stem cell research*. Retrieved August 8, 2012, from http://www.hinxtongroup.org/au_trans_cs.html.

Kass, L. R. (2002). *Human cloning and human dignity: An ethical inquiry. Report of the President's Council on Bioethics*. New York: PublicAffairs.

Mooney, C. (2006). *The republican war on science*. New York: Basic Books.

Murugan, V. (2009). Embryonic stem cell research: A decade of debate from Bush to Obama. *Yale Journal of Biology and Medicine, 82*(3), 101–103.

National Institutes of Health Guidelines for Human Stem Cell Research, 74 Fed. Reg. 32,170 (July 7, 2009).

Nebraska prohibits state funding of hESC research absolutely: Neb. Rev. Stat. § 71-7606 (2007).

Salter, B. (2007). Patenting, morality and human embryonic stem cell science: Bioethics and cultural politics in Europe. *Regenerative Medicine, 2*(3), 301–311.

Secretary's Advisory Committee on Genetics, Health, and Society. (2009). *The integration of genetic technologies into health care and public health*. http://oba.od.nih.gov/SACGHS/sacghs_documents.html.

Shulman, S. (2006). *Undermining science: Suppression and distortion in the Bush Administration*. Los Angeles: University of California Press.

Speech of Hon Lamar Smith of Texas, 112 *Cong. Rec.* E1182 (Jun. 23, 2011).

Stout, D., & Harris, G. (2009, March 6). Obama reversing stem cell limits Bush imposed (pp. 2). New York Times.

Torremans, P. L. (2010). Legal problems raised by patents on human stem cell—Based interventions. In K. Hug & G. Hermerén (Eds.), *Translational stem cell research* (pp. 287–307). New York: Springer.

Vrtovec, K. T., & Scott, C. T. (2008). Patenting pluripotence: The next battle for stem cell intellectual property. *Nature Biotechnology, 26*(4), 393–395.

Cases

American Fruit Growers v. Brogdex Co., 283 U.S. 1 (1931).

Funk Bros. Seed Co. v. Kalo Inoculant Co., 333 U.S. 127 (1948).

In re Bergy, 596 F.2d 952 (C.C.P.A. 1979).

Diamond v. Chakrabarty, 447 U.S. 303 (1980).

Association for Molecular Pathologists et al. v. U.S. Patent and Trademark Office et al., 702 F. Supp.2d 181 (S.D.N.Y. 2010), affirmed in part, reversed in part, 653 F.3d 1329 (Fed. Cir. 2011), *certiorari* granted, decision vacated and remanded, 2012 U.S. LEXIS 2356 (U.S. Mar. 26, 2012).

Opinion of the Advocate General, Case C-34/10, Brustle v. Greenpeace, 10 March 2011.
Oliver Brüstle v. Greenpeace eV, European Court of Justice C-34/10, OCTOBER 18, 2011
Sherley v. Sebelius, 644 F.3d 388 (2011).
Mayo Collaborative Services Inc. v. Prometheus Laboratories, Inc., 132 S.Ct. 1289 (2012).
Association for Molecular Pathologists et al. v. U.S. Patent and Trademark Office et al., 698 F.3d 1303 (Fed. Cir. 2012).

Statutes

The Leahy Smith America Invents Act, Pub. Law 112–29, 125 STAT. 284. September 16, 2011
Consolidated Appropriations Act, 2010, Pub. Law 111–117, 123 Stat 3034
Germany, *Bundestag*, *Stem Cell Act of 2002* (Stammzellgesetz), Bundesgesetzblatt [Federal Law Gazette] 2002, Part I, no. 42, p. 2277, June 29, 2002, §1-1.
35 U.S.C. §112, 1st paragraph (1952).
35 U.S.C. § 271(a) (1952).
Neb. Rev. Stat. § 71-7606 (2007).
Md. Code, Com. Law §10-434 (2008)
Mass. Gen. Laws ch. 111L, §12J (2008).
Euro Parliament and Council: Directive 98/44/EC "Directive on the legal protection of biotechnological invention. Brussels, Belgium, July 6, 1998.

Chapter 5
Repositories for Sharing Human Data in Stem Cell Research

Pilar N. Ossorio

5.1 Introduction

As stem cell science moves forward, researchers will increasingly use high-throughput methods and "-omics" to characterize cells and to decipher the molecular choreography of development and differentiation. Researchers will use genomics, epigenomics, proteomics, transcriptomics, metabolomics, reactomics, and probably other "omics" yet to be invented. Omics and other high-throughput research approaches are extremely data intensive. For example, the 1000 Genomes Project—an international project to generate a catalog of human genetic variation (The 1000 Genomes Project Consortium 2010)—generated more DNA sequence data in its first 6 months than all of the sequence data accumulated in GenBank over its previous 21 years of existence (Benson et al. 2012; Stein 2010). The explosion of research data is illustrated by a recent study by personnel from the journal Science, which found that 20 % of scientists surveyed regularly used data sets exceeding one gigabyte and about 7 % regularly used data sets exceeding one terabyte (Anonymous 2011).

The January 2012 special database issue of Nucleic Acids Research described 92 new online molecular biology databases (new since the January 2011 issue), and the journal's online catalog listed 1,380 databases and data repositories (Galperin and Fernandez-Suarez 2012).[1] Well-known and widely used databases include GenBank (Benson et al. 2012), Ensembl (Flicek et al. 2012) and the European Nucleotide Archive (Amid et al. 2012), which provide submission, search, and

[1] As of July, 2012, the Nucleic Acids Research online database catalog could be found at http://www.oxfordjournals.org/nar/database/a/.

P.N. Ossorio, Ph.D., J.D. (✉)
University of Wisconsin, School of Law and School of Medicine and Public Health,
Morgridge Institute for Research, Madison, WI, USA
e-mail: pnossorio@facstaff.wisc.edu

L.F. Hogle (ed.), *Regenerative Medicine Ethics: Governing Research and Knowledge Practices*, DOI 10.1007/978-1-4614-9062-3_5,

download services for DNA sequence data. The Universal Protein Resource (UniProt) serves protein data (The UniProt Consortium 2012), and the Gene Expression Atlas, the Gene Expression Omnibus (GEO) (Sayers et al. 2012), and ArrayExpress serve gene expression data. There are repositories for data on epigenomes, pathogen genomes, yeast genomes, viral genomes, bacterial toxins, small molecule binding sites, NMR results, macromolecular interactions, secondary structural motifs in RNA three-dimensional structures, and many more biological molecules and phenomena.[2]

Most twentieth-century databases were designed for archiving, sharing, and comparing scientific research results. The data elements consisted of single gene or protein sequences, or other small data types. Increasingly, twenty-first-century data repositories are designed as discovery resources—tools for generating new results. Such repositories aim to take full advantage of the transformations that digital technologies have brought to biomedical research; their data elements include whole genome analyses, whole cell proteome analyses, and other similarly large data elements, often accompanied by substantial demographic and phenotypic data. To increase the utility of large data sets, some repositories facilitate the sharing of raw or lightly computed data from next-generation platforms. The database of Genotypes and Phenotypes (dbGaP) (Mailman et al. 2007), the European Genome-Phenome Archive (EGA) (Leinonen et al. 2011), and the NCBI and European Sequence Read Archives (SRA) (Kodama et al. 2012) are discovery resources that facilitate sharing of raw data. Data from such repositories can be analyzed by numerous research groups who ask different scientific questions and/or use different analysis algorithms, and the repository data can be combined with other data sets to create new research materials.

Over the past decade, science funding agencies and journals have set forth policies to promote data sharing. Stem cell scientists will be expected to contribute to general purpose data sharing repositories such as dbGaP or an SRA, and are already beginning to create stem-cell-specific repositories. For instance, the Stem Cell Discovery Engine is an online, curated database of cancer stem cell experiments coupled to the Galaxy analysis software package (Sui et al. 2012). The Stem Cell Omics Repository is a queryable database of large-scale human embryonic stem cell (hES cell) and induced pluripotent stem cell (iPS cell) data, including proteome and protein phosphorylation data (Phanstiel et al. 2011). The Stem Cell Lineage Database (SCLD) contains data on cell-type-specific gene expression, cell lineage maps, and stem cell differentiation protocols for humans and mice (Hemphill et al. 2011). These repositories vary in the degree to which they support robust data sharing and discovery research, but stem cell research will no doubt move further and further in the direction of sharing raw or lightly computed data.

[2] For a comprehensive review of new repositories and databases for the biological sciences, see the January 2012 database supplement of Nucleic Acids Research (volume 40, issue D1), and for a discussion of the difficulties of creating, maintaining, and analyzing large data sets, see the February 2011 issue of Science.

This chapter will provide a brief description of funders' current data sharing requirements and the justificatory underpinnings for those requirements. It will then note some ethical tensions that arise when the data being shared derive from human beings. Finally, it will describe existing governance mechanisms designed to resolve these ethical tensions.

Throughout this chapter, people whose data are in repositories will often be referred to as "sources" or "data sources" instead of "human subjects" or "research participants." Research using coded or de-identified data (or biological specimens) from repositories often does not meet regulatory definitions of research on a human subject, so describing data sources as human subjects could lead to confusion regarding oversight requirements (Ossorio 2011). Furthermore, data sources may not know their data are in a repository and usually have no interaction with scientists who use repository data in research. For these reasons, it seems misleading to describe most data sources as "research participants."

5.2 Data Sharing Requirements

5.2.1 Norms and Justifications for Data Sharing

Science funding agencies have identified several reasons to promote data sharing. Such sharing is consistent with scientific and academic norms of open inquiry and collective knowledge production within an epistemic community. Sharing may enable a variety of scientists to test alternative hypotheses or alternative analytic methods on one data set, thereby increasing the data's value. Funders and policy makers prefer that some data be analyzed by several independent groups because it is increasingly unlikely that any single research group will have the capacity to extract or develop all of the information available in a very large, complex data set. In addition, some data are effectively unique because they are so expensive and difficult to generate that they will not be replicated in the foreseeable future. Society, the publics who often fund research, and scientists themselves would potentially lose out on valuable knowledge if the data generators were the only people who could analyze unique data. Finally, data sharing permits scientists to combine different data sets to form new discovery tools.

Sharing allows research groups to validate one another's results. It is impossible to validate published conclusions without access to the original data. "Without access to research data, we are asked to accept findings as an act of faith, rather than as a scientific conclusion" (Berman 2002). Data sharing also provides some safeguards against scientific misconduct. When the data must be widely available for others to use, researchers are less likely to fabricate or falsify data or results.

Overall, funders, science policy experts, and many researchers have become strong advocates of data sharing because it allows society to leverage the value of each data set produced.

5.2.2 Funders and Journals May Require Data Sharing

The National Institutes of Health currently requires a data sharing plan for any research project with $500,000 or more of direct costs in any year. This requirement applies to "final research data," defined as: "factual material necessary to document, support, and validate research findings" (Office of Extramural Research 2003). Final research data means something more than the summary or aggregate data included in published journal articles. Final research data must become available no later than the acceptance for publication of the main findings generated from the final data. The form in which the data become available and the processes for distributing, curating, and storing them are features to be spelled out in the researchers' sharing plan. In rare circumstances, the plan may state that the research group will not share its data or will only share under limited circumstances. NIH's default position is that data should be shared, so a plan to limit sharing must have a very strong justification.

NIH guidance documents strongly encourage the deposition of all genome-wide association data (including accompanying phenotype and other "metadata") into a federal repository. dbGaP is usually the designated repository (National Institutes of Health 2007). Depositing data into a designated repository is not the only means by which those data may be shared, and deposition requirements are not intended to prevent researchers from sharing through their own websites or through other sharing platforms.

NIH sometimes imposes project-specific data sharing requirements, particularly for large projects that will generate unique or difficult-to-replicate data. Projects such as the 1000 Genomes Project (The 1000 Genomes Project Consortium 2010) or the Human Microbiome Project (Turnbaugh et al. 2007) have project-specific governance, including data sharing policies. Large, multisite, international projects may have a data coordinating center to facilitate data standardization, quality control, and sharing.

In the United Kingdom, the Wellcome Trust "expects all of its funded researchers to maximize the availability of research data with as few restrictions as possible" (Wellcome Trust 2010). The Trust requires grant applicants whose projects will likely generate a data output of significant value to the wider scientific community to include a data management and sharing plan in their proposals. Genome Canada has very similar policies (Genome Canada 2008), and in January, 2011, 17 major science funding agencies from 7 countries signed a joint statement supporting the norm of widespread data sharing to promote and improve public health (Wellcome Trust 2011).

In addition to funder requirements, some journals require that DNA and protein sequences, expression array data, or other specified data be deposited in an appropriate data repository as a condition of publication. Some journals simply require that published authors make their data available to others on request, and some journals strongly encourage data sharing but do not require it as a condition of publication.

A recent study found that of 351 papers covered by a journal's policy requiring or suggesting data sharing, 59 % did not make data available through a repository or other online source (Alsheikh-Ali et al. 2011). (The papers at issue came from a wide spectrum of scientific fields, not only the biological sciences.) At nearly the same time, another survey of scientists found that 67 % agreed or strongly agreed with the statement "Lack of access to data generated by other researchers or institutions is a major impediment to the progress of science" (Tenopir et al. 2011)! Clearly, many scientists want more data sharing, and at the same time many are falling short of the data sharing ideal.

5.2.3 Prepublication Data Release

Both the NIH and the Wellcome Trust have long required *prepublication* release of data from resource-generating projects. Resource-generating projects are designed from the outset to produce data, biological materials, and analytical tools for widespread use by a large group of scientists. The Human Genome Project (HGP) was the canonical example of a resource-generating project, and its inception drove promulgation of prepublication data sharing principles (Collins et al. 2003; Contreras 2011). These principles were articulated in a series of consensus meetings that included data producers, data users, science funders, journal editors, and a few people with expertise in law and ethics. The 1996 International Strategy Meeting on Human Genome Sequencing gave rise to the "Bermuda Principles" for open access genomics and prepublication data release (Kaye et al. 2009). These principles were reaffirmed in the Fort Lauderdale Agreement of 2003 and extended in the Toronto Statement of 2009 (Toronto International Data Release Workshop 2009). A 2008 meeting of proteomics researchers in Amsterdam—the International Summit on Proteomics Data Release and Sharing Policy—led to the articulation of similar data sharing principles for proteome research (Rodriguez et al. 2009).

Attendees at the 2009 meeting in Toronto "endorsed the value of rapid prepublication data release for large reference data sets in biology and medicine that have broad utility and agreed that prepublication data release should go beyond genomics and proteomics studies to other data sets…." The Toronto Statement extended the principle of rapid, prepublication data release beyond resource-generating projects. Instead, the Statement calls for prepublication data release when a project is "large scale" (defined as requiring significant resources over time), when the data have broad utility, when the data generated could be viewed as "a reference data set," and when the relevant scientific community accepts that prepublication data release is appropriate (Toronto International Data Release Workshop 2009). Although there is no explicitly agreed upon definition of a reference data set, generally such data can serve as the primary comparator for many subsequent studies. For instance, the HGP produced a reference sequence against which subsequent human genome data could be compared, and the Haplotype Mapping Project created a catalog of common linkage disequilibrium blocks from three populations that has been used a comparator in hundreds of studies (The International HapMap Consortium 2007).

The Toronto Statement set forth responsibilities of data producers, data users, funders, and journals. Making explicit such responsibilities helps assure that data sharing does not overly burden any one constituency and that all view it as fair. Data sharing, particularly prepublication data sharing, raises issues of trust and fairness between data producers and data users. Data producers worry that users will scoop them and publish using the producers' data before the producers can publish. For this reason, some projects for which funders require prepublication data release have a "publication embargo" (Contreras 2011). This is a period during which the data are available, but users cannot submit a publication without permission of the data producers. During the embargo period, users are encouraged to collaborate with data producers so that all can be included as authors on submitted papers.

5.2.4 *Barriers to Data Sharing*

Some commentators have argued that data and materials sharing among stem cell scientists is hindered by factors such as strategic behavior by scientists, the rapid pace of scientific progress, the ethical complexities of using human cell lines and data, and intellectual property rights (patents or the desire to obtain patents) (Mathews et al. 2011; The Hinxton Group 2010). Numerous other bioscience fields have faced similar barriers, and to varying extents, overcome them.

At times, data sharing has seemed more an aspiration than a scientific practice. In a survey of geneticists conducted in the early 2000s, 12 % reported denying at least one request for data underlying results published in the preceding 3 years (Campbell et al. 2002). The primary reason for not sharing was that it required too much effort. The second most common reason for withholding data was that respondents were protecting their trainees' or their own opportunities to publish. Other surveys have found scientists reluctant to share data because of privacy and confidentiality risks for data sources (discussed in Sect. 5.3 below), and the desire to comply with their institutions' policies on invention disclosure and intellectual property (Tenopir et al. 2011).

Whether scientists actually share data is strongly influenced by whether their funders and home institutions encourage and support the activity. Funders who require data sharing are paying to support the preparation, annotation, and archiving of data. They are paying for creation and maintenance of data repositories and for the development of computational tools to ease and standardize data deposition pipelines. Funders are paying for conferences at which scientists develop data ontologies and come to consensus on the basic information that should be included with certain types of repository submissions.[3] As funders become more enamored of data sharing, they are helping to lower barriers to the activity.

A great deal of stem cell research in the United States is funded by NIH and will be subject to its data sharing requirements and to the data sharing culture NIH is encouraging among its funded scientists. However, research on hES cells receives less public funding than most other types of biology, so scientists who derive and study these cells may receive less support for hES-cell-specific data sharing activities, and strong data sharing norms may be slower to emerge.

Private sector funders of biomedical research sometimes view data sharing as in their best interests. Several private sector entities, including large pharmaceutical companies such as Merck and Pfizer, have contributed funding and other support to projects designed to create publicly available community resources for research (Contreras 2011). For example, ten pharmaceutical and biotechnology companies collaborated with the Wellcome Trust to form and fund the nonprofit SNP Consortium, which mapped single nucleotide polymorphisms in the human genome and submitted the mapping data to a public database—dbSNP (Sherry et al. 2001; Woodman 1999). Thus, even private sector stem cell funders might come to view some data sharing as being in their interests and might be willing to support the activity.

5.2.5 *Patents and Data Sharing*

The complex patent landscape of human pluripotent stem cells has been singled out as a particular barrier to data sharing among stem cell scientists (Mathews et al. 2011; The Hinxton Group 2010). The relationship of data sharing to patenting is complex. Prepublication data sharing was intended, in part, to preclude patenting. The HGP's prepublication data sharing policy was intended to put data into the public domain and preempt the opportunity for patents on genomic DNA sequence (gDNA) (Collins et al. 2003; Contreras 2011). Unlike mRNA or cDNA sequences, which might be used to express proteins with medical, agricultural, or other

[3] One of the earliest and best-known standards for information sharing is MIAME—minimum information about a microarray experiment (Brazma et al. 2001). As microarray gene expression studies became a widely used source of genome-scale data in the life sciences, scientists discovered that much of the microarray data being generated were inaccessible or unusable by most of the interested scientific community. Microarray data accompanying publications were typically reported on the authors' website, using a variety of formats. There was no consensus on what annotation was necessary, and the data often were not annotated. Data generally did not include indicators of reliability and quality. As the scientific community moved to create repositories for microarray data sharing (such as GEO and ArrayExpress), a consensus group was formed to suggest what types of information should be included in the repositories. The MIAME standards were the consensus group's attempt to specify minimum information that must be included in a repository to make the data interpretable and usable to the broader scientific community. Since MIAME, other consensus groups have attempted to specify minimum information standards for other types of data.

economic value, HGP policy makers considered gDNA sequence "precompetitive information" and felt that the balance of interests favored putting such precompetitive information in the public domain. Data from high-throughput molecular experiments on stem cells might also be viewed as precompetitive information, in the economic, if not the professional, sense.

However, making data publicly available *does not preclude patenting of discoveries made using those data*. Making a cell's whole genome genotype publicly available could prevent another researcher from obtaining a patent on that genotype, but it would not prevent somebody from patenting the finding of an association between a SNP in that genotype and a particular drug response,[4] even if that association was found by using the publicly available genotype (in conjunction with several thousand other cell's genotypes). Thus, putting data, especially raw data, in the public domain does not preclude all patenting.

Finally, even if a research result is patented, the patent holder could still share it without restriction, or with minimal restriction. This could be done using an open-source-type license (Rai 2005). The possibility or existence of patents does not necessarily preclude data sharing.

5.3 Anonymity and Risk in Human Data Sharing

When the data to be shared derive from nonhumans, the ethical and legal issues revolve around intellectual property, fairness between data producers and users, and the development of good governance structures. Among other things, good governance structures assure data quality and consistency, help researchers negotiate disagreements, and help assure that all applicable rules or laws have been followed by scientists who generate and share data.[5] However, when the data derive from humans, the issues are much more complex and raise questions of risk minimization and data source autonomy. This chapter cannot address all of the relevant ethical issues and will focus on the question of risks to data sources.

The risks from disclosing sensitive human data are informational. Informational risks can be broken down into potential harms to individuals if information linked back to them is used to harm them or their loved ones, and risks to individuals or groups if information about the group is linked back to members of the group and

[4] Associations between a genetic or biochemical marker and a health outcome have been patentable in the United States for several decades. However, in 2012, the US Supreme Court held that a patent on a method of calibrating a drug's dose by assessing levels of the drug's metabolites in a person's blood was not patentable because it was a law of nature (Mayo Collaborative Services v. Prometheus Laboratories, Inc. 2012). This case and others currently before the courts may substantially limit researchers' and firms' opportunities to patent correlations between biomarkers and health states or treatment outcomes. See, also, Noonan (this volume).

[5] When the project involves an international collaboration, differences in national laws concerning data security and transfer, national security, and privacy can make research governance extremely difficult (Zink and Silman 2008).

used in harmful ways. Group harm from biomedical research has been discussed by others, and the context of data sharing through a research repository does not substantially change those analyses (Foster 1998; Foster et al. 1997; Foster and Sharp 2007; Juengst 1998; Sharp and Foster 2000; Weijer et al. 1999), so this chapter will focus on the risk that disclosure of individual information might harm data sources. To assess this risk, we must consider what types of information are and might be included in data sharing repositories, how likely it is that information from a repository could be disclosed to unauthorized persons and linked back to a particular person, and what would happen if such information were linked back to a particular person.

5.3.1 Many Data Repositories Contain Highly Sensitive Information

dbGaP is probably the largest repository for sharing raw or lightly computed data. As of July, 2012, it contained data from 312 different genome studies (National Center for Biotechnology Information 2012). Depending on the study, the genomic data could be individual whole genome genotypes, whole exome sequences, whole genome sequences, or other large-scale, individual-level sequence. In the future, data sharing repositories will probably include several types of molecular data, such as whole genome and expression array or epigenome data from the same individual.

A data sharing repository for promoting genome research will also contain individual-level data pertaining to any trait or traits under study. The non-genomic data associated with each genotype or sequence are often referred to as "metadata." Depending on the study, the metadata could consist of medical information, exposure information, behavioral information, and demographic information. Currently, some of the sensitive, individual-level information in dbGaP includes sexual behavior, smoking behavior, diagnoses of mental illness, microbial communities associated with a woman's vagina, and information on personal hygiene habits. In many cases, people consider these types of information private and personal and do not share them with friends, family members, or employers. At the very least, such information could be embarrassing; at worst it could lead to ostracization or job loss.

A data sharing resource for hES cell research might include information about the embryo progenitors' health and demographics, possibly whole genome data from each progenitor, as well as whole genome data from the hES cells. A resource for sharing iPS cell or adult stem cell data likely would include whole genome data from the cell line, perhaps other "omic characterization" of the cells, along with the cell source's medical history (including information about any relevant diagnoses), sociodemographic characteristics, and other information relevant to particular studies.

5.3.2 Data Disclosures and Potential Harms

When considering what risks a sharing repository might impose on data sources, one must consider both the molecular data and the metadata. A genotype or large-scale sequence will disclose many health risks. Particularly if the data source is young enough that such risks may not yet have materialized, she may not want them known to unauthorized persons. Some health conditions are stigmatizing or embarrassing, and many people do not share even the potential of developing these conditions for fear that the information might be used against them by friends, future spouses, employers, or others.

The United States has legislation, the Genetic Information Nondiscrimination Act (GINA), which aims to prevent employers or health insurers from seeking to obtain genetic information or from using any genetic information they inadvertently obtain in ways that disadvantage people (Genetic Information Nondiscrimination Act 2008). However, other countries have no comparable legislation, and many data sets in sharing repositories contain information pertaining to people from several countries. Even in the United States, GINA does not prevent insurers from obtaining and using genetic information to deny or to price life, disability, long-term care, or other insurance coverage. Furthermore, proving that one has been the subject of illegal genetic discrimination by an employer or health insurer can be difficult, so some people may not want to depend on GINA's protections. And finally, people may not want certain information known to friends, coworkers, or community members, even if that information is not used for illegal discrimination. A person's friends or relatives could stigmatize her and exclude her from the social activities or communities she values.

In summary, even with GINA, inadvertent disclosure of a person's whole genome information to unauthorized parties could lead to a variety of economic, social, or psychological harms.

A person may not always want to know her own genetic risks. When Jim Watson had his genome sequenced, he chose to make nearly all of the sequence fully public. However, he did not want to know his own genetic risk for Alzheimer's disease, so he chose to have certain portions of his genome withheld from the public and from himself (Wheeler et al. 2008). Inadvertent disclosure of unwanted genetic information to the data source is a psychological risk. [6]

[6] Shortly after Dr. Watson's genome sequence was published, Nyholt et al. published an article describing how one could infer Dr. Watson's genetic risk for Alzheimer's using linkage disequilibrium between genetic markers in the published sequence and the redacted portion of Dr. Watson's genome (Nyholt et al. 2009). The Nyholt authors demonstrated that their method worked by using it to infer the Alzheimer's risk alleles in Craig Venter's published sequence. As a consequence of this work, Dr. Watson and the scientists who sequenced his genome redacted an additional two megabases of his sequence around genes associated with Alzheimer's disease risk. Nyholt et al. point out that as the scientific community's knowledge of genetic risk and linkage disequilibrium increases, it will become more difficult to protect a person from unwanted information disclosure by withholding a portion of her otherwise public genome.

Perhaps the most significant informational risks come from the possible disclosure of metadata. Even simple data such as blood pressure or measures of C-reactive protein could be used by an insurer to increase premiums or deny benefits (at least in countries without universal health care, and where the laws permit denial of coverage based on preexisting conditions or rate increases based on one's state of health). GINA does not cover health insurers' or employers' use of non-genetic information (Hudson et al. 2008). When research participants initially share sensitive information with researchers, they may do so only because the participants have a high degree of confidence that the information will remain confidential and will be used only for important scientific research. Unauthorized disclosures of sensitive personal information may cause social, economic, or psychological harm to data sources and may undermine the public's trust in researchers and the research enterprise.

Stem cell researchers, research participants, ethicists, and social scientists working together will be necessary to comprehend the informational risks that might arise during stem cell research and data sharing. The following hypothetical may provide some insights, however. Suppose a couple has experienced three miscarriages and has no biological children. The couple attempts in vitro fertilization, and in the process, scientists determine that each embryo has a chromosomal anomaly incompatible with survival to birth. The couple agrees that each will have a genome analysis done and that researchers may make stem cells from the embryos. Researchers believe these cells will be helpful in understanding the biology of early human development, and they conduct extensive molecular analyses of the hES cells derived from the couple's embryos, including genotyping and gene expression profiling. Researchers also collect extensive family histories, medical histories, and exposure information from each parent-progenitor. What types of risks would be present if such information was inadvertently disclosed?

In the hypothetical above, disclosing the couple's decision allowing embryonic stem cells to be created from their embryos could make them targets of anti-stem cell protesters and could get them ostracized from their church or shunned by family members. It could cause one or both to be fired from their jobs. Furthermore, the couple might not have shared their history of miscarriages with many people. Having that information find its way into unauthorized hands and connected back to them could raise a host of traumatic memories. These are but some of the potential harms that could materialize if information in a hES cell data sharing repository could be traced back to particular individuals.

5.3.3 Protecting Sources from Informational Risks: Anonymization

Traditionally, biomedical researchers have protected study participants and data sources from informational risks by anonymizing or coding the data. Scientists have operationalized the concept of anonymity by stripping all explicit identifiers from the data (and also from biological materials such as blood or cells). Types of identifiers that are typically removed include names, social security numbers, telephone numbers, addresses (including email addresses), medical record numbers, URLs, and photos. Data or research materials from which all identifiers have been stripped are referred to as anonymized.

Rather than fully anonymizing data (or biological materials), explicit identifiers may be replaced by a code. The decryption key needed for re-identifying the source is then stored securely and separately from the coded data, and not transmitted to data users. Coding, rather than full anonymization, is necessary for some research designs. For instance, coding is necessary for the duration of all longitudinal studies because coded data can be updated. When human-derived stem cells are used in clinical trials or medical treatment, regulators will probably require robust and frequently updated information about the cell sources. Thus, coded rather than fully anonymized cells and information likely would be used.

Researchers, some ethicists, and some research participants/data sources view coded data as preferable to fully anonymized data (or biological materials) (National Bioethics Advisory Commission 1999). If the data are coded, then the code can be broken when doing so would benefit sources or researchers. For instance, the source can be re-identified and contacted if the researchers discover clinically or reproductively relevant information in the source's data (Wolf et al. 2008, 2012). Recontact is also desirable to researchers who seek consent for additional research projects or who are conducting quality control on a data set. Bioethicists, scientists, and policy makers are currently engaged in a contentious, wide-ranging debate concerning whether, when, and how to recontact research participants and data sources.

Data sharing repositories typically contain coded data. Inclusion of explicit identifiers usually would not be necessary for research and would unacceptably increase risks to data sources. Furthermore, in the United States (and probably elsewhere), both the creation and use of a repository that included explicit identifiers would receive a high degree of regulatory scrutiny (Ossorio 2011).[7]

5.3.4 Genomes, Individuating Data, and the Loss of Anonymity

Recently, commentators have questioned whether people's identities can be adequately protected when their data are in complex data sets such as those in sharing repositories (Benitez and Malin 2010; Heeney et al. 2011; Homer et al. 2008; Lowrance and Collins 2007; Malin 2005a, b; Malin et al. 2010; McGuire and Gibbs 2006; Yeniterzi et al. 2010). For numerous reasons, many commentators believe that

[7] In the United States, both the regulations for the protection of human participants in research (the "Common Rule," codified at 45 CFR Part 46) and the Health Insurance Portability and Accountability Act's Privacy Rule (codified at 45 CFR Parts 160 and 164) would place oversight and consent requirements on a repository that included explicit identifiers.

there is a rapidly increasing probability that sensitive information in a research repository could be connected back to a particular individual even though explicit identifiers were not included in the repository.

The anonymity of research data sets containing whole genome information is particularly suspect, because such information is person-unique (unless the DNA is from a monozygotic twin) (Lowrance 2002, 2006). Approximately 70 single nucleotide polymorphism markers (SNPs) randomly distributed across the genome are enough to individuate most people in the world (Lin et al. 2005), and researchers have estimated that 100,000 contiguous nucleotides would be enough to individuate the vast majority of people. In addition, one's DNA sequence is stable throughout one's lifetime (with the caveat that particular cells or cell lineages may develop somatic mutations or rearrangements) and relatively easily replicated by many people. Any type of stable, replicable, person-unique element in a data set increases the risk that the source will be re-identified. Genomic information individuates people and carries an intrinsic risk of identifying the person from whom it derived.

Genomic information from hES cells will relate to the embryo from which the cells derived and will not individuate a living person. However, such genomic information might be enough to identify the embryo's progenitors. For iPS and other stem cells and cell lines, the genetic information contained therein will be the same as that of the person from whom the cells were derived (unless the cell line has acquired chromosomal rearrangements or mutations after the cells were cultured).

Of course, even information that individuates a person will not identify her unless it can be linked back to her (Malin et al. 2010).

5.3.5 Re-identification Science

Anonymity and identifiability are not discrete categories; they define the ends of a continuum concerning the ease with which a particular person can be "got at." Rather than speaking categorically about identifiable or anonymized data, researchers and ethicists ought to speak of the probability that an unauthorized person could identify a source in a data set or the degree of anonymity provided by particular identity-obscuring strategies. The degree of identifiability/anonymity is not a simple trait or characteristic intrinsic to data; rather, the degree of identifiability/anonymity depends on the context in which the data are used and the knowledge of the person attempting re-identification. For example, a person's medical record number can only re-identify her if the recipient of that number also has access to her medical record; the number may be meaningless in other contexts. Yet, the probability that an unauthorized person could connect a named individual to information in a medical record is great enough that medical record numbers have long been treated as identifiers for regulatory purposes.

Three conditions must be present to re-identify somebody—there must be individuating information in the data, there must be a resource for obtaining contact information for data sources, and there must be a mechanism to relate the anonymized but individuating data to the contact information (Malin et al. 2010). Unfortunately, there are a growing number of situations in which these three conditions are present.

There are already many situations in which de-identified biomedical information can be re-identified without hacking or breaking into private information systems. One type of re-identification attack, sometimes called the "Netflix attack" or the "Netflix-type attack," involves linking information from a large, anonymized data set to information from another data set that contains contact information for people who are sources for both data sets. Netflix is associated with this type of attack because in 2006 it publicly released millions of anonymized user records to facilitate crowd sourcing of research on algorithms to predict user film ratings (Ohm 2010). However, two scientists from the university of Texas re-identified many individuals by linking the anonymized Netflix data with information from the Internet Movie Rating Database, which included raters' names (Narayanan and Shmatikov 2006). Patterns of movie preferences were used to connect information between the two databases and therefore to connect the identifying information in the movie rating database to the anonymized, individual-level data in the Netflix database.

An earlier example of the Netflix-type attack occurred in 1996, when it took a researcher less than a week to re-identify the medical records of the Massachusetts governor by merging publicly available, anonymized hospital discharge records with voter registration records from MA (Sweeny 1996). Subsequent research indicates that between 63 and 87 % of the US population could be uniquely identified using a combination of date of birth, gender, and residential zip code and connecting that information to voter records or other publicly available sources containing explicitly identifying information. Data mining has become a great deal more sophisticated since 1996; more algorithms have become available for analyzing and combining data sets.

Clearly, in the Netflix-type attack the probability that a person in an anonymized data set may be re-identified depends on the reasonable availability of other data pertaining to the data source. The more databases contain a person's information, the more likely she is to be re-identified.

People whose data are in research repositories are also included in a rapidly growing number of other databases compiled and held by both public and private entities. A great deal of medical information that is or will be contained in research data sets also resides in databases held by firms such as insurers, fitness companies, wellness programs, and employers. Many of these non-research databases contain identifiable information. For marketers or insurers, the point of data collection and compilation is often to construct individual profiles, so identifiers will be linked to or included in the data. There is an increasing likelihood that biomedical research data could be combined with data held by outside entities to re-identify research data sources. Once information about a person from two or more databases is combined, not only may the data source be re-identified, but a rich profile of her may also be created.

DNA sequence or genotype information will be particularly useful for linking databases. Genomic information is now being compiled by genealogy companies, dating websites, purveyors of dietary supplements, paternity testing firms, cosmetic companies, law enforcement agencies, immigration authorities, and many others. In the future, one's DNA sequence or large segments of it may be included in one's medical records at more than one health-care provider. Because DNA sequence is (usually) person-unique, it is particularly useful for combining an individual's information from two or more databases.

Whether other omic data, such as epigenome, proteome, or expression array data, will be person-unique, reproducible, and stable over time has yet to be determined. Scientists participating in the Human Microbiome Project recently announced at an invitational meeting that a person's gut microbiome may be unique and stable over time. This preliminary observation must be verified; however, it raises the prospect that person-unique data may arise in unexpected situations.

Other types of re-identification attacks may also put data sources at risk (Malin 2005a, b, 2006; Malin et al. 2010). Space limitations prevent a thorough description here, but suffice it to say that the more data and the more different types of data are contained in a data set, the higher the risk of re-identification. Data that describe features or characteristics of a person that are documented in multiple environments, or are publicly available, create a higher risk of re-identification (Malin et al. 2010).

5.3.6 Data Aggregation Does Not Necessarily Mask Source's Identities

Another means of minimizing informational risk has been to aggregate data prior to publishing or otherwise sharing it. Unfortunately, for some types of data this technique is not adequate. In 2008, Homer et al. published an article demonstrating that one can pick an individual genotype out of complex, aggregate data even when the genotype in question represented less than 0.1 % of the total information in the aggregate (Homer et al. 2008; Sankararaman et al. 2009). Homer et al. provided both a computational model and an experimental demonstration of their approach, and researchers at NIH later verified Homer's findings. To identify an individual from aggregate genotype data, one must possess an identified copy of the genotype; however, as discussed above, people's genomic information now can be found in a variety of databases. There is an increasing likelihood that somebody will have a source's genotype and could therefore pick it out of aggregated research results. Homer's re-identification method has been called a "pool attack" (Malin et al. 2010).

Harm from a pool attack would arise if identifying the person as part of an aggregate data set disclosed that she had a condition or trait under study, such as depression, schizophrenia, or a sexually transmitted disease. Another type of harm could materialize if the research database contained sensitive metadata not originally available to the holder of the identified genotype.

Homer's findings resulted in both the NIH and the Wellcome Trust removing aggregate genome information from public websites. However, Schadt and colleagues recently demonstrated that gene expression data are also susceptible to the pool attack, but some unrestricted websites still distribute such data (Schadt et al. 2012). Gene expression data can be used to predict a person's genotype at some loci that control nearby gene expression (cis eQTLs), and under some conditions, the predicted genotypes can be matched to a person's known genotype with a high degree of accuracy. Thus, with known genotypes, a person could identify individuals whose gene expression data are in a research data set. Several repositories and individual investigators have for years posted gene expression data on websites that are available to the public without restriction. Some continue to do so.

Schadt and colleagues noted that gene expression data contain a wealth of information beyond that pertaining to the medical condition under study. Gene expression levels are strongly correlated with the existence of cancers, body mass index, lipid levels, glucose levels, age, sex, and other traits. Thus, if a person is identified as the source of data in a gene expression study, a great deal of information about her could be revealed.

5.3.7 Creating Future Repositories

When stem cell scientists create research repositories, they ought to consider the degree to which necessary data elements, or a combination of data elements within a data set, will be person-unique. Creators of repositories also ought to consider the data environment and whether (or in what form) elements in their research data sets are likely included in data held by other entities. Rather than assume that a de-identification strategy adequately decreases risks to data sources, researchers ought to perform explicit re-identification risk assessments while designing data repositories and prior to sharing particular data sets (Benitez and Malin 2010). Projects are currently underway to develop computational tools that will help researchers determine re-identification risks.

As discussed in the introduction, regulators and scientists view a great deal of research using coded or de-identified data (or biological specimens) as falling outside of the Common Rule's definition of human subjects research. Research involves human subjects if an investigator obtains "identifiable private information" about a living individual (Department of Health and Human Services 2005). Identifiable means that "the identity of the subject is or may readily be ascertained by the investigator or associated with the information" (Department of Health and Human Services 2005). How great does the risk of re-identification have to be before data should be treated as "identifiable enough" that societies will require ethics oversight when those data are used in research? What strategies and oversight processes are most effective and efficient for minimizing informational risks associated with complex, high-dimensional data? Policy makers, ethicists, and scientists are currently debating these questions.

5.4 Governing Data Sharing and Protecting Research Participants

As researchers and policy makers have recognized that it may be difficult to truly anonymize research data sets, they have developed complex repository governance mechanisms to balance society's interests in promoting the progress of science through data sharing with society's and individual source's interests in protecting research participants/data sources from informational harm.

Data sharing on a large scale involves a triad—the data producers who conduct research that generates large data sets, the repository, and data users (other than the producers) who download or access the data to conduct research. Users of repository data are sometimes described as conducting "secondary research." This term refers to the timing of their research and not to its quality or importance.

Sharing data not derived from humans usually involves broadcasting the data to a fully public website in a standardized format that is useful for large numbers of scientists. However, when the data derive from human beings and pose informational risks, researchers and policy makers have chosen alternative sharing mechanisms, often referred to as "controlled access." Controlled access repositories place restrictions on who can use the data, limit the purposes for which data may be used, and require that data users comply with various data security measures. The NIH genome-wide association study (GWAS) rules, instantiated in dbGaP policies, specify the most fully elaborated and formally documented controlled access process (National Institutes of Health 2007; Ossorio 2011). NIH uses the GWAS rules, with slight additions and modifications, for other required data sharing, such as sharing of whole exome or whole genome sequence data. Most controlled access repositories have rules similar to the GWAS rules.

The GWAS rules aim to provide several layers of protection for data sources. The first layer is implemented by the data producers' home institutions. These institutions must certify that the data were generated in compliance with all applicable laws and regulations, including regulations for the protection of human subjects in research (National Institutes of Health 2007). In addition, the institution must state that deposition of the data into dbGaP, and subsequent sharing, is consistent with the original consent under which the data were generated (if there was consent for the research).[8] The data-producer institution must not deposit information that could explicitly identify data sources in the repository, and must certify that the investigator's plan for de-identifying the data complies with professional standards. Prior to submitting data to the repository, the producer's institution must evaluate those data

[8] Sometimes, molecular analysis is conducted using leftover clinical specimens for which no consent for research was obtained or for which the purported consent constituted a one-line authorization to use excess tissue in research. Unconsented research on specimens originally collected for clinical treatment or diagnoses is allowed under the Common Rule. Institutions differ as to whether data from such studies may be deposited in a repository for broad data sharing. Some institutions require researchers to contact or recontact individuals and obtain consent for data sharing (Ludman et al. 2010).

and "consider the risks to individuals, their families and groups or populations…" (National Institutes of Health 2007). While NIH strongly encourages data sharing, it leaves open the possibility that a data producer's institution would determine that the risks to individuals, families, or groups sometimes outweigh the social benefits of data sharing. In the first instance, the onus is on data producers' institutions to determine whether any data set may be shared.

A second layer of protection for data sources is implemented by the repository. The repository must apply prescribed data security standards to protect the data it holds. Although not mandated by a formal rule, the repository also conducts quality checks on incoming data and, among other things, ensures that explicit identifiers were not inadvertently included. The repository screens prospective users to determine that they will use the data for a legitimate research purpose and in compliance with any restrictions that may apply. The repository specifies data security standards and other rules that will legally bind data users.

The third level of protection is implemented by data users. Users, their home institutions, and sometimes an associated information technology specialist sign a contract with the repository agreeing to abide by rules designed to protect data sources. Under the NIH GWAS rules, users and their institutions agree to implement prescribed data security measures, to abide by any restrictions on the types of experiments that can be done using the data, and to use the data only for the purpose set forth in their "data access request." Users promise not to re-identify any data source and not to sell any elements of a data set. Users may only share a data set, or part of it, with other people who are named on the user's data access request (National Institutes of Health 2007).

5.4.1 *Impacts of Controlled Access Governance*

Controlled access clearly and intentionally creates impediments to data sharing and likely diminishes the number of people who will use a data set. Whether controlled access rules adequately protect data sources remains to be seen. To date, no harmful data security breach has occurred. In a series of interviews conducted from mid-2010 to February of 2012, the author of this chapter discovered that most data users work diligently to protect the data but that a variety of "minor" rule violations occur (Ossorio 2011). These violations involve sharing data with somebody who is not approved to use it, but who could have been approved and who is often added to the data access request after the fact; using data for an unapproved type of analysis; and using data past the date on which a researcher has agreed to destroy them.

Many data users believe the "transaction costs" imposed by controlled access prevent dilettantes from using the data but do not really slow the advancement of knowledge. Such users believe that scientists who have a serious purpose for using the data, and who are likely to publish papers based on the data, will obtain the data even though they must endure a sometimes aggravating access process that takes anywhere from several weeks to several months. In the future, policy makers and scientists may develop less burdensome, less bureaucratic mechanisms for protecting data sources.

One incidental protection for data sources is that the data sets at issue are often so large that they cannot (yet) be downloaded to or used on laptop or tablet computers. They cannot be stored on "thumb drives" or other small, portable devices. The immense size of these data sets places practical restriction on their mobility. Such restrictions may not exist in the future.

5.5 Conclusion

Stem cell scientists will increasingly be expected to deposit raw or lightly computed data into sharing repositories. When the data derive from human cells, those repositories likely will operate according to a controlled access process, at least for the foreseeable future. As stem-cell-specific data repositories are created for data from humans, scientists and IT specialists should consider the informational risks associated with the types of data to be shared. They should be aware that stripping extrinsic identifiers or aggregating data may not sufficiently diminish informational risks. A controlled access process will likely be necessary for some stem-cell-specific repositories.

References

Alsheikh-Ali, A. A., Qureshi, W., Al-Mallah, M. H., & Ionnidis, J. P. A. (2011). Public availability of published research data in high impact journals. *PLoS One, 6*, e24357. doi:10.1371/journal.pone.0024357.

Amid, C., Birney, E., Bower, L., Cerdeno-Tarraga, A., Cheng, Y., et al. (2012). Major submissions tool developments at the European nucleotide archive. *Nucleic Acids Research, 40*, D43–D47.

Anonymous. (2011). Challenges and opportunities. *Science, 331*, 692–693.

Benitez, K., & Malin, B. (2010). Evaluating re-identification risks with respect to the HIPAA Privacy Rule. *Journal of the American Medical Information Association, 17*, 169–177.

Benson, D. A., Karsch-Mizrachi, I., Clark, K., Lipman, D., Ostell, J., et al. (2012). Genbank. *Nucleic Acids Research, 40*, D48–D53.

Berman, J. J. (2002). Confidentiality issues for medical data miners. *Artificial Intelligence in Medicine, 26*, 25–36.

Brazma, A., Hingamp, P., Quackenbush, J., Sherlock, G., Spellman, P., et al. (2001). Minimum information about a microarray experiment (MIAME)—Toward standards for microarray data. *Nature Genetics, 29*, 365–371.

Campbell, E. G., Clarridge, B. R., Gokhale, M., Birenbaum, L., Hilgartner, S., et al. (2002). Data withholding in academic genetics: Evidence from a national survey. *Journal of the American Medical Association, 287*, 473–480.

Collins, F. S., Morgan, M., & Patrinos, A. (2003). The human genome project: Lessons from large-scale biology. *Science, 300*, 286–290.

Contreras, J. (2011). Bermuda's legacy: Policy, patents, and the design of the genome commons. *Minnesota Journal of Law, Science and Technology, 12*, 61–125.

Department of Health and Human Services. (2005). Protection of Human Subjects, 45 Code of Federal Regulations, Part 46.102(f).

Flicek, P., Amode, M. R., Barrell, D., Beal, K., Brent, S., et al. (2012). Ensembl 2012. *Nucleic Acids Research, 40*, D84–D90.

Foster, M. W. (1998). Model agreement for genetic research. *American Journal of Human Genetics, 63*, 696–702.

Foster, M. W., Eisenbraun, A. J., & Carter, T. H. (1997). Communal discourse as a supplement to informed consent for genetic research. *Nature Genetics, 17*, 277–279.

Foster, M. W., & Sharp, R. R. (2007). Share and share alike: Deciding how to distribute the scientific and social benefits of genomic data. *Nature Reviews Genetics*. doi:10.1038/nrg2124.

Galperin, M. Y., & Fernandez-Suarez, X. M. (2012). The 2012 nucleic acids research issue and the online molecular biology database collection. *Nucleic Acids Research, 40*, D1–D8.

Genome Canada. (2008). *Data release and resource sharing*. Retrieved June 25, 2012, from http://www.genomecanada.ca/medias/PDF/EN/DataReleaseandResourceSharingPolicy.pdf.

Genetic Information Nondiscrimination Act of 2008, Pub. Law 110-233, 122 Stat. 881. (2008).

Heeney, C., Hawkins, N., Jd, V., Boddington, P., & Kaye, J. (2011). Assessing the privacy risks of data sharing in genomics. *Public Health Genomics, 14*, 17–25.

Hemphill, E. E., Dharia, A. P., Lee, C., Jakuba, C. M., Gibson, J. D., et al. (2011). Scld: A stem cell lineage database for the annotation of cell types and developmental lineages. *Nucleic Acids Research, 39*, D525–D533.

Homer, N., Szelinger, S., Redman, M., Duggan, D., Tembe, W., et al. (2008). Resolving individuals contributing trace amounts of DNA to highly complex mixtures using high-density snp genotyping microarrays. *PLoS Genetics, 4*(e1000167), 1000161–1000169.

Hudson, K. L., Holohan, M. K., & Collins, F. S. (2008). Keeping pace with the times—The genetic information nondiscrimination act of 2008. *New England Journal of Medicine, 358*, 2661–2663.

Juengst, E. T. (1998). Groups as gatekeepers to genomic research: Conceptually confusing, morally hazardous, and practically useless. *Kennedy Institute of Ethics Journal, 8*, 183–200.

Kaye, J., Heeney, C., Hawkins, N., de Vries, J., & Boddington, P. (2009). Data sharing in genomics—Re-shaping scientific practice. *Nature Reviews Genetics, 10*, 331–335.

Kodama, Y., Shumway, M., Leinonen, R., & International Nucleotide Sequence Database Collaboration. (2012). The sequence read archive: Explosive growth of sequencing data. *Nucleic Acids Research, 40*, D54–D56.

Leinonen, R., Akhtar, R., Birney, E., Bower, L., Cerdeno-Tarraga, A., et al. (2011). The European nucleotide archive. *Nucleic Acids Research, 39*, D28–D31.

Lin, Z., Owen, A. B., & Altman, R. B. (2005). Genomic research and human subject privacy. *Science, 305*, 183.

Lowrance, W. W. (2002). *Learning from experience: Privacy and the secondary use of data in health research*. London: The Nuffield Trust.

Lowrance, W. W. (2006). *Privacy, confidentiality and identifiability in genomic research. Workship on Privacy, Confidentiality and Identifiability in Genomic Research. Oct. 3–4*. Bathesda: National Human Genome Research Institute.

Lowrance, W. W., & Collins, F. S. (2007). Identifiability in genomic research. *Science, 317*, 600.

Ludman, E. J., Fullerton, S. M., Spangler, L., Trinidad, S. B., Fujii, M. M., et al. (2010). Glad you asked: Participants' opinions of re-consent for dbGaP data submission. *Journal of Empirical Research on Human Research Ethics, 5*, 9–16.

Mailman, M. D., Feolo, M., Jin, Y., Kimura, M., Tryka, K., et al. (2007). The NCBI dbGaP database of genotypes and phenotypes. *Nature Genetics, 39*, 1181–1186.

Malin, B. (2005). Betrayed by my shadow: Learning data identity via trail matching. *Journal of Privacy Technology,* 20050609001.

Malin, B. (2005b). An evaluation of the current state of genomic data privacy protection technology and a roadmap for the future. *Journal of the American Medical Information Association, 12*, 28–34.

Malin, B. (2006). Re-identification of familial database records. In *AMIA 2006 symposium proceedings* (pp. 525–528).

Malin, B., Karp, D., & Scheuermann, R. H. (2010). Technical and policy approaches to balancing patient privacy and data sharing in clinical and translational research. *Journal of Investigative Medicine, 58*, 11–18.

Mathews, D. J. H., Graff, G. D., Saha, K., & Winickoff, D. E. (2011). Access to stem cells and data: Persons, property rights, and scientific progress. *Science, 331*, 725–727.

Mayo Collaborative Services v. Prometheus Laboratories, Inc., 132 S. Ct. 1289. (2012).

McGuire, A. L., & Gibbs, R. A. (2006). No longer de-identified. *Science, 312*, 370–371.

Narayanan, A., & Shmatikov, V. (2006). *Robust de-anonymization of large sparse datasets*. Retrieved July 12, 2012, from http://www.Cs.Utexas.Edu/~shmat/shmat_oak08netflix.Pdf.

National Bioethics Advisory Commission. (1999). *Research involving human biological materials: Ethical issues and policy guidance, Volume I*. Rockville: National Bioethics Advisory Commission.

National Center for Biotechnology Information. (2012). *DbGaP*. Washington, DC: National LIbrary of Medicine.

National Institutes of Health. (2007). Policy for sharing of data obtained in NIH supported or conducted genome-wide association studies (GWAS). *Federal Register, 72*, 49290–49297.

Nyholt, D., Yu, C.-E., & Visscher, P. (2009). On Jim Watsons APOE status: Genetic information is hard to hide. *European Journal of Human Genetics, 17*, 147–150.

Office of Extramural Research. (2003). *Nih data sharing policy and implementation guidance*. Retrieved February 12, 2012, from http://grants.Nih.Gov/grants/policy/data_sharing/data_sharing_guidance.Htm.

Ohm, P. (2010). Broken promises of privacy: Responding to the surprising failure of anonymization. *UCLA Law Review, 57*, 1701–1777.

Ossorio, P. (2011). Bodies of data: Genomic data and bioscience data sharing. *Social Research, 78*, 907–932.

Phanstiel, D. H., Brumbaugh, J., Wenger, C. D., Tian, S., Probasco, M. D., et al. (2011). Proteomic and phosphoproteomic comparison of human ES and iPS cells. *Nature Methods, 8*, 821–827.

Rai, A. K. (2005). "Open and collaborative" research: A new model for biomedicine. In R. W. Hahn (Ed.), *Intellectual property rights in frontier industries*. Washington, DC: AEI Press.

Rodriguez, H., Snyder, M., Uhlen, M., Andrews, P., Beavis, R., et al. (2009). Recommendations from the 2008 international summit on proteomics data release and sharing policy: The Amsterdam principles. *Journal of Proteome Research, 8*, 3689–3692.

Sankararaman, S., Obozinski, G., Jordan, M. I., & Halperin, E. (2009). Genome privacy and limits of individual detection in a pool. *Nature Genetics, 41*, 966–967.

Sayers, E. W., Barrett, T., Benson, D. A., Bolton, E., Bryant, S. H., et al. (2012). Database resources of the National Center for Biotechnology Information. *Nucleic Acids Research, 40*, D13–D25.

Schadt, E. E., Woo, S., & Hao, K. (2012). Bayesian method to predict individual SNP genotypes from gene expression data. *Nature Genetics, 44*, 603–609.

Sharp, R. R., & Foster, M. W. (2000). Involving study populations in the review of genetic research. *Journal of Law, Medicine & Ethics, 28*, 41–51.

Sherry, S. T., Ward, M., Kholodov, M., Baker, J., Phan, L., et al. (2001). dbSNP: The NCBI database of genetic variation. *Nucleic Acids Research, 29*, 308–311.

Stein, L. D. (2010). The case for cloud computing in genome informatics. *Genome Biology, 11*, 207.

Sui, S. J. H., Begley, K., Reilly, D., Chapman, B., McGovern, R., et al. (2012). The stem cell discovery engine: An integrated repository and analysis system for cancer stem cell comparisons. *Nucleic Acids Research, 40*, D984–D991.

Sweeny, L. (1996). *Uniqueness of simple demographics in the u.S. Population*. Working Paper LIDAP-WP4. Data Privacy Lab, Carnegie Mellon University, Pittsburgh, PA.

Tenopir, C., Allard, S., Douglass, K., Aydinoglu, A. U., Wu, L., et al. (2011). Data sharing by scientists: Practices and perceptions. *PLoS One, 6*, e21101. doi:10.1371/journal.pone.0021101.

The 1000 Genomes Project Consortium. (2010). A map of human genome variation from population-scale sequencing. *Nature, 467*, 1061–1073.

The Hinxton Group. (2010). *Statement on policies and practices governing data and materials sharing and intellectual property in stem cell science*. Retrieved February 12, 2012, from http://www.Hinxtongroup.Org/consensus_hg10_final.Pdf.

The International HapMap Consortium. (2007). A second generation human haplotype map of over 3.1 million SNPs. *Nature, 449*, 851–861.

The UniProt Consortium. (2012). Reorganizing the protein space at the universal protein resource (UniProt). *Nucleic Acids Research, 40*, D71–D75.

Toronto International Data Release Workshop. (2009). Prepublication data sharing. *Nature, 461*, 168–170.

Turnbaugh, P. J., Ley, R. E., Hamady, M., Fraser-Liggett, C. M., Knight, R., et al. (2007). The human microbiome project. *Nature, 449*, 804–810.

U.S. Agency for Healthcare Research and Quality, Bill and Melinda Gates Foundation (U.S.), U.S. Centers for Disease Control, Doris Duke Charitable Foundation (U.S.), U.S. Health Resources and Services Administration, Hewlett Foundation (U.S.), U.S. National Institutes of Health, U.S. Substance Abuse and Mental Health Services Administration, Canadian Institutes of Health Research, Deutsche Forschungsgemeinschaft, Economic and Social Research Council (UK), Medical Research Council (UK), Wellcome Trust (UK), Health Research Council of New Zealand, INSERM (FR), National Health and Medical Research Council (Australia), The World Bank. The list of signatories can be found at: http://www.wellcome.ac.uk/About-us/Policy/Spotlight-issues/Data-sharing/Public-health-and-epidemiology/Signatories-to-the-joint-statement/index.htm

Weijer, C., Goldsand, G., & Emanuel, E. J. (1999). Protecting communities in research: Current guidelines and limits of extrapolation. *Nature Genetics, 23*, 275–280.

Wellcome Trust. (2010). *Data management and sharing*. Retrieved June 15, 2012, from http://www.Wellcome.Ac.Uk/about-us/policy/spotlight-issues/data-sharing/data-management-and-sharing/index.Htm.

Wellcome Trust.(2011). Sharing research data to improve public health: Full joint statement by funders of health researcher. Retrieved July 8, 2012, from http://www.Wellcome.Ac.Uk/about-us/policy/spotlight-issues/data-sharing/public-health-and-epidemiology/wtdv030690.Htm.

Wheeler, D. A., Srinivasan, M., Egholm, M., Shen, Y., Cen, L., et al. (2008). The complete genome of an individual by massively parallel DNA sequencing. *Nature, 452*, 872–877.

Wolf, S. M., Crock, B. N., Van Ness, B., Lawrenz, F., Kahn, J. P., et al. (2012). Managing incidental findings and research results in genomic research involving biobanks and archived datasets. *Genetics in Medicine, 14*, 361–384.

Wolf, S. M., Paradise, J., Nelson, C. A., Kahn, J. P., & Lawrenz, F. (2008). Managing incidental findings in human subjects research: Analysis and recommendations. *Journal of Law, Medicine & Ethics, 36*, 219–248.

Woodman, R. (1999). Wellcome Trust and drug giants fund gene marker database. *British Medical Journal, 318*, 1093.

Yeniterzi, R., Aberdeen, J., Bayer, S., Wellner, B., Hirschman, L., et al. (2010). Effects of personal identifier resynthesis on clinical text de-identification. *Journal of the American Medical Information Association, 17*, 159–168.

Zink, A., & Silman, A. (2008). Ethical and legal constraints on data sharing between countries in multinational epidemiological studies in Europe: Report from a joint workshop of the European League Against Rheumatism standing committee on epidemiology with the "AutoCure" Project. *Annals of Rheumatic Disease, 67*, 1041–1043.

Part II
Research Integrity: Updates for Regenerative Medicine Researchers

Chapter 6
Beyond the Checkboxes: Research Integrity for Regenerative Medicine Researchers

Jessica Von Reyn, Amritava Das, and Linda F. Hogle

6.1 Introduction

Responsible conduct of research (RCR) also referred to as "good scientific practice" or "scientific integrity," can be defined as the practice of scientific investigation with integrity, incorporating professional standards and ethical principles. Broadly speaking, the term refers collectively to human subjects protections, animal protections, laboratory safety guides, and the more recent attention to professionalism issues such as misconduct, conflicts of interest (COIs), and credit for work done. Allowing unethical practices—even those committed unintentionally or through carelessness—can have serious consequences. Misrepresented data or biased findings might unwittingly be used by others as the basis for further experiments, resulting in wasted effort and resources. The reputation of scientists and science in general may be damaged, not to mention the possibility of considerable harm to research subjects. Misconduct by an investigator may also affect mentoring relationships within a lab, and unfairly harm lab members' careers. Standards for ethical conduct provide researchers with a touchstone for their own research practices and help to establish trust among researchers, the public, funders, and the subjects of research. Research ethics guidance is also needed in order to function within increasingly complex collaborative arrangements, some of which cross institutional or national borders with different policies. Researchers in any scientific endeavor should thus

J.Von Reyn
School of Law, University of Wisconsin-Madison, Madison, WI 53706, USA

A. Das
Department of Chemical and Biological Engineering, 1415 Engineering Dr., Madison, WI 53703, University of Wisconsin-Madison USA

L.F. Hogle (✉)
University of Wisconsin—Madison, 1300 University Avenue, 1135 Medical Sciences Building, Madison, WI 53706, USA
e-mail: lfhogle@wisc.edu

L.F. Hogle (ed.), *Regenerative Medicine Ethics: Governing Research and Knowledge Practices*, DOI 10.1007/978-1-4614-9062-3_6,

be familiar with the essential components of research integrity and know how to recognize poor practices both in their own and others' work.

The Office of Research Integrity (ORI) of the U.S. Department of Health and Human Services (DHHS) has identified nine essential components of RCR. Most research institutions require scientists to receive formal training on these components and a number of online and published resources have been created for this purpose. However, the content of these resources is rarely updated, and may be too general to provide sufficient guidance for the real-world needs of scientists in constantly changing research environments. Also, some of the case studies seem to be designed for the needs of clinical researchers or others who work under different conditions than those found in cell biology, engineering, or bioinformatics research. As a result, RCR training modules may not be perceived as helpful and may too easily be dismissed by some as simply a bureaucratic exercise—a matter of "checking the boxes" required to get on with their research. Yet both experienced researchers and those new to the field will face situations which are ambiguous or pose novel ethical challenges due to the nature of the research; they need tools to discern and solve potential problems. Answers may not easily be found within the boxes, however. Researchers must know where to go for advice and be proactive in seeking help for complicated situations. At the same time, policy makers and ethicists must stay abreast of commonly encountered circumstances in contemporary bioscience and periodically review the boxes and their contents. Novel areas of research such as embryonic stem cells, nanomedicine, synthetic biology, and gene editing may pose new ethical questions, as will some of the newer social arrangements involved in conducting science, such as interdisciplinary, large-scale, or academic-industry collaborations. It may or may not be necessary to create additional or different oversight guidelines and administrative mechanisms in each case. Regardless, oversight should be measured and appropriate for the variety of research activities that exist, and it should be applied systematically; that is, all involved in research should know how to recognize potential problems and understand responsibilities for oversight. We invite readers (researchers and policy makers alike) to "think beyond the checkboxes" of existing research integrity components to consider how best to responsibly conduct research.

To this end, this chapter offers a brief general overview of core components of the RCR as identified by the ORI, including proposed changes, some of which have been hotly debated. We provide illustrations salient for emerging fields of science including examples of gray areas and those in which sets of rules may appear to conflict. Many of the examples we provide apply specifically to stem cell-related research, but the information provided is relevant to broader areas of cell and molecular science, biomedical engineering, bioinformatics, and more. The chapter begins with a review of some basics of RCR administrative structures and responsibilities, including information specifically relevant to stem cell research. This will be helpful to those new to research ethics issues, as it provides background for the principles underlying current ethics infrastructures plus information on where to turn for guidance. Readers more interested in specific applications may wish to skip to Sect. 6.3, which unpacks the RCR boxes described in Sect. 6.2 and provides case illustrations. The chapter ends by expanding upon some areas which have been less discussed, but which may have important implications going forward (Sect. 6.4). As with many areas of ethical, legal, and professional practice guidelines, there are ambiguities and limitations to existing ways of thinking about what constitutes integrity in research. The chapter is not intended to be an exhaustive resource; instead, sources for more detailed information are provided throughout the text. Readers are encouraged to consult requirements and key personnel identified by their own universities or organizations for additional information specific to their institutions.

6.2 The Basics of Scientific Integrity: Who Defines the Checkboxes?

This section outlines key organizations involved in the administration of research integrity. In most cases, more than one agency is involved in research integrity efforts, with varying degrees of authority to implement policy or punish violators. Some have legislative authority, while others have the power to enforce compliance through the ability to sanction researchers or withhold funding.

Oversight systems, agencies, and specific rules for RCR evolve over time and in response to perceived needs. A public controversy may draw attention to a problem, leading to tightened oversight or new rules, or lobbying efforts might lead to loosened oversight or weakening of agencies' roles. New technologies can spark debates about whether existing oversight is sufficient or should be modified, or if an overhaul is needed. Recent examples include nanotechnology and synthetic biology (Presidential Commission 2010; Fatehi et al. 2012). RCR guidelines and oversight systems, then, can be historically, politically, and socially contingent, and there was not a single, overarching logic from whence existing rules sprang. While there is considerable consensus in the identification of common guiding principles, attempts to coordinate RCR activities across institutions have been met with mixed success.

6.2.1 Administrative Organization and Oversight Authority

In the United States, the ORI of the DHHS was established to promote integrity in biomedical and behavioral research supported by the U.S. Public Health Service (PHS).[1] The office sets policies, procedures, and regulations regarding research misconduct, reviews intramural and extramural policies and procedures, and promotes RCR through educational activities. It also assists institutions dealing with misconduct and reviews misconduct investigations. Should misconduct be found, the ORI recommends administrative actions to the DHHS, and assists the Office of

[1] The ORI, created in 1992, is under the auspices of the Office of Public Health and Science (a part of the Department of Health and Human Services) and has jurisdiction over ten offices and agencies, including the NIH, CDC, and FDA.

the General Counsel in preparing cases on appeal when warranted. There are some jurisdictional reporting differences among major federal funding agencies: for example, the NIH's guidelines are enforced by the ORI whereas the Office of the Inspector General has jurisdiction over NSF researchers. Also, the ORI, which evolved from concerns about clinical research and human subject abuses, oversees activities related to health and medicine. RCR for other areas of science and engineering is now addressed under a specific provision of the America COMPETES Act (2007, renewed in 2010), which grew out of a concern about economic competitiveness and was intended to stimulate science research, education, and entrepreneurship. The Act is largely concerned with activities supporting commerce (including enhancing the education pipeline for a trained workforce); hence, questions around COI are of greater concern than the human subjects focus of ORI. The Act provides targeted funding for specific areas of research, but does not fund mandated RCR activities or articulate how RCR training should be carried out.

Most private research funders and all US public (governmental) sources of funds require principal investigators, students, and key personnel to have instruction in RCR as a condition of granting funds.[2] Institutions receiving funds are responsible for maintaining documentation detailing appropriate participation in the training, and are charged with assuring that general criteria are met.[3] Both the NIH and NSF require that each grant proposal includes a plan for RCR training and oversight to undergraduate students, graduate students, and postdoctoral researchers participating in the proposed research project, but leave the specifics to individual institutions.

---

[2] Notice number NOT-OD-10-019 (November 24, 2009) applied the policy to the following NIH programs: D43, D71, F05, F31, F32, F33, F34, F37, F38, K01, K02, K05, K07, K08, D12, K18, K22, K23, K24, K25, K26, K30, K99/R00, KL1, KL2, R25, R36, T15, T32, T34, T35, T36, T37, T90/R90, TL1, TU2, and U2R.DY. The NSF requirement of RCR training was formalized in the America Creating Opportunities to Meaningfully Promote Excellence in Technology, Education, and Science America Act, or the so-called America COMPETES Act (42 U.S.C. 18600–1, Sec 7009). It applies to all full research proposals submitted after January 4, 2010.

[3] The NIH criteria include the following: (a) the training must not be solely through online instruction; there is value in discussions and instruction with research faculty and other grantees. (b) The training should cover nine areas of responsible conduct, including conflict of interest, human and animal subject practices, mentorship responsibilities and relationships, collaborative research, peer review, data acquisition and laboratory tools, research misconduct, authorship and publication, and the scientist's role in society (see Table 6.2). (c) The institution's research facility should take a mentoring role in providing both formal and informal instruction. (d) An effective program should contain at least 8 h of instruction, although a complete semester of programming will result in more significant learning. (e) Researchers are responsible for ongoing training throughout their careers: during graduate school (generally through a seminar-type course), at the early investigator level (through discussions with mentors and possible formal programming), and as senior fellows as career award recipients (as mentors, lecturers, and discussion leaders). The NSF has funded several projects to provide resources for institutions and relevant personnel including the Ethics Collaborative Online Resource Environment (CORE) digital library, at http://nationalethicscenter.org/ and the Online Ethics Center for Engineering and Research. Six broad categories are covered, including emerging technologies, environment, safety and sustainability, employment and legal issues, professional practice, responsible research, and diversity issues (found at http://www.onlineethics.org).

Internationally, the expectations and definitions of responsible and ethical research vary greatly. In Europe, each nation has its own individual guidelines for its scientists and their international collaborators, but 30 nations have signed the European Code of Conduct for Research Integrity, a 2010 creation of the European Science Foundation (ESF) and All European Academies (ALLEA).[4] In China, the ethics of emerging research are governed by the country's Ministry of Health and the Ministry of Science and Technology, which enacted the nation's policy of stem cell research in 2003 (Zhang 2012). Like China, Japan has issued national guidelines to guide stem cell research pursuant to the recommendations of its Council of Science and Technology Policy and Expert Panel on Bioethics (Normile 2009; Kawakami et al. 2010).

The Second World Conference on Research Integrity resulted in a document outlining basic principles of research integrity, known as the "Singapore Statement on Research Integrity."[5] The document is not an official representation of participants' policies and acknowledges differences among countries, but recognizes that there are fundamental values on which research should be based. These include honesty, accountability, fairness, and stewardship, and added a dimension of social responsibility often missing in individual countries' policies. For example, institutions and individuals are advised of their ethical obligations to weigh societal benefits against risks inherent in their work, and the researchers' responsibilities to limit professional comments to their recognized expertise when engaged in public discussions about research. Canada created a new framework for research integrity in 2011. Integrity as defined in this policy encompasses rigorous and careful analysis, commitment to the dissemination of research results, and the importance of the responsible use of public funds.[6] The Singapore Statement and Canadian Council documents are significant because they establish a framework grounded in value-based principles, rather than procedural norms, as the US system does.

While there have been attempts to harmonize governance across national jurisdictions, there are locally specific characteristics which researchers should know about, particularly when conducting collaborative research in those countries. This is particularly important for research areas considered to be ethically sensitive, such as the use of embryos in research, the creation of novel life forms (including genetically altered organisms or chimeras), or dual-use technologies (e.g., those which could be used for benefit or harm, such as bioweapons). A compilation of International Human Research Protections, with more than 1,000 guidelines,

---

[4] The European Science Foundation reports on research integrity can be found here: ESF.org/activities/mo-for a/research-integrity.html.

[5] The document is so named because the meeting was held in Singapore in July of 2010. The statement can be found at: Retrieved September 16, 2013.

[6] The Canadian Tri-agency framework for Responsible Conduct of Research covers the three major funding agencies (Canadian Institutes of Health Research (CIHR), the Natural Sciences and Engineering Research Council (NSERC), and the Social Sciences and Humanities Research Council (SSHRC)) and can be found here: http://www.rcr.ethics.gc.ca/eng/policy-politique/framework-cadre/.

regulations, and laws from 101 countries lists key organizations and covers issues ranging from data sharing to specific issues related to human testing of devices, drugs, genetic technologies, and, most recently, stem cells.[7] While some RCR components have general consensus about what is unethical behavior (e.g., falsifying data), the widely varying ethical and legal circumstances regarding human embryonic stem cell research (hESC) across countries have made harmonization of policies difficult for some aspects of the research (e.g., the use of human tissue).[8]

6.2.2 *Policies Specific to Embryonic Stem Cell Research*

In the absence of any clear federal laws governing stem cell research, both governmental and nongovernmental organizations have formulated their own recommendations for ethical practice.[9] The differences and duplications among them create a confusing and uncertain legal and ethical space for stem cell practice (Levine 2011).

In the United States, the National Academy of Sciences (NAS) drafted a set of guidelines covering human subjects concerns (including informed consent and payment of embryo and gamete donors) and derivation practices (National Academy of Sciences 2005). The original guidelines were amended in 2010 to clarify areas of overlap with newer US federal regulations, and to add or modify specific recommendations.[10] The NAS is a private, not-for-profit scientific advisory body and has no enforcement power: compliance with its recommended guidelines is voluntary and applies only to US researchers. However, the NAS has significant status as a scientific authority, and most institutions have adopted the recommendations, recognizing the benefits of a unified approach with a codified set of expectations for all stem cell researchers. The 23 original components, written in 2005, have served as a guide for other organizations and other countries, although there are some significant variances. It is important to note that individual institutions reviewed each guideline, and then accepted, rejected, or modified them in a way which they deemed most suitable for their own institutions.[11] There was general agreement that the Guidelines should

[7] The 2013 list can be accessed at http://www.hhs.gov/ohrp/international/intlcompilation/intlcomp2013.pdf.pdf.

[8] For example, research involving the use of embryos is statutorily forbidden in Germany, permitted under country-specific conditions in the UK, France, and Switzerland, and subject to (relatively) less regulation in China (Gottweis 2002, 2006; McMahon et al. 2010; Isasi & Knoppers 2006).

[9] There is often some confusion about types of policies, and there may be inconsistencies among them: guidelines are not considered binding by the authorities issuing them, whereas regulations issued by governmental agencies such as the NIH and the CIRM (California Institute for Regenerative Medicine) are binding, and may make funding conditional on compliance.

[10] The original 2005 Guidelines can be found at: http://www.nap.edu/catalog/11278.html and the 2010 Guidelines at http://www.nap.edu/catalog.php?record_id=12923.

[11] For example, while it is generally agreed that embryo donors should not be paid for their donations due to moral concerns about the nature of embryos, Guideline #16 also bans payment for gamete donors. Many institutions did not adopt the regulation, since such donors are regularly remunerated for other purposes, and often at market rates, rather than merely covering the donor's basic expenses.

be adopted as completely as possible, to lend consistency and transparency to the process and to aid interoperability where multi-institutional or multinational collaborations exist. Most institutions have instituted specific training based on the NAS guidelines and mandate compliance for both investigators and their collaborators.

The Guidelines also recommend the establishment of embryonic stem cell research oversight (ESCRO) committees at sites conducting embryonic stem cell research.[12] These committees are tasked with reviewing protocols utilizing embryonic stem cells. ESCROs function in addition to, and in coordination with, existing local institutional oversight bodies, such as Institutional Review Boards (IRBs) which govern human subjects protections, Animal Care and Use Committees (IACUCs) which oversee the use of animals in research, and, where necessary, biosafety committees. It is worth noting that a number of universities had created advisory bodies prior to the existence of NAS recommendations, and in some states, this function is mandated apart from the NAS. The work of ESCRO has changed over time, from establishing institutional policy to a more bureaucratic protocol-reviewing function. Greeley argues that ESCROs may have evolved past their original purpose; that is, since many of the protocol review issues have to do with creating new hESC lines, possible insertion of hESCs into humans, or creating chimeras, protocol reviews could instead be done by IRBs and IACUCs (which already deal with human and animal subjects) (Greely 2013). However, he acknowledges that unforeseen policy issues often arise (such as the advent of induced pluripotent stem cells (iPSCs), or new, controversial techniques), which may require review by some form of advisory body. The question is whether this should be done at a national, international, or local institutional level.

The NAS provides guidelines for which kinds of research protocols must be reviewed and what types of research are impermissible (see Table 6.1). These are based on a general agreement about what was morally permissible to do with materials derived from human embryos, and in view of US federal laws regarding the use of human embryos in research.[13] Any derivation of a new line must be reviewed by an ESCRO, but purely in vitro research using existing lines listed in the U.S. NIH registry may receive expedited review, as these lines have been previously reviewed for compliance with federal policies regarding their derivation and provenance. According to the guidelines, any research where the identity of donors is ascertainable is not permissible due to privacy concerns, as is the in vitro culture of any intact human embryo for more than 14 days due to moral concerns about the developing human.[14] Because of the potential for moral concerns about mixing

[12] Some institutions have decided to expand their purview beyond embryonic stem cell research, anticipating the need to review iPS cells and possibly more, and thus have dropped the "E" from the acronym.

[13] There is no complete consensus on the moral status of embryos or what should be morally permissible, but the NAS reached its conclusions based on consultation with legal and ethical authorities and went forward with the best understanding of what could be agreed upon at that time.

[14] Fourteen days of development is the point at which the primitive streak forms in humans. Beyond this point, the neural system begins to appear, and eventually, sensation or awareness might be possible. For this reason, 14 days is chosen as the point beyond which research should not continue (Downs 2008).

Table 6.1 National Academy of Sciences protocol review guidelines

Full review required	• Derivation of new lines • Introduction of hESCs into nonhuman animals at any embryonic, fetal, or postnatal stage • Introduction of hPSCs into nonhuman animals at any embryonic, fetal, or postnatal stage if an expected effect is that human cells will integrate into the central nervous system, testes, or ovaries of the animal • Research in which personally identifiable information about donors of embryos, gametes, or somatic cells from which cells were derived is readily ascertainable by the investigator
Expedited review	• Purely in vitro research using preexisting coded lines
Impermissible research	• In vitro culture of any intact human embryo for >14 days • Research in which hPSCs are introduced into nonhuman primate preimplantation embryos, or in which pluripotent stem cells from any species are introduced into human preimplantation embryos • Research involving breeding of animals where the introduction of hPSCs may contribute to the germ line • Research where identity of donors is ascertainable

human and nonhuman organisms, it is also not permissible to introduce pluripotent stem cells into nonhuman primate preimplantation embryos or pluripotent stem cells from any species into a human preimplantation embryo. Research involving the creation of chimeras in which the introduction of pluripotent stem cells may contribute to the germ line or neural tissues is also forbidden. Section 6.3.2 contains further discussion on the use of chimeras.

Importantly, the Guidelines require that provenance of the embryos and their donors be documented; that is, it must be known that consent was properly obtained for each embryo from which a line was derived. If lines are acquired from other institutions, documentation must be obtained from that institution.

The NAS is not the only body setting guidelines for stem cell researchers. Some states in the United States have their own guidelines, some of which are more or less restrictive than national guidelines and some of which have since been superseded by the 2009 NIH Guidelines (Lomax and Stayn 2008). Most notably, the California Institute for Regenerative Medicine (CIRM) has created a distinct set of guidelines for research that they fund, as well as a unique review process, which includes participation from representatives of the state's public. Under the CIRM guidelines, hESC lines derived from any embryo donated with voluntary and informed consent are permissible, whereas the NIH forbids the use of hESC lines created from embryos for purposes other than reproduction.

In addition to public governmental sources, a number of professional associations have either added to their own existing recommendations for good practice or have created new ones. In particular, the International Society for Stem Cell Research (ISSCR), a voluntary international consortium of stem cell organizations, also evaluated ethical issues and set guidelines[15] (ISSCR 2006). The guidelines are similar to the NAS regarding reproductive cloning, excessive compensation regarding egg donation, donor privacy, and banning the culturing of embryos beyond 14 days. Unlike the NAS, which recommends an institutional oversight mechanism (that is, the formation of ESCRO committees), the ISSCR simply recommends that institutions provide a mechanism through which the review occurs effectively, impartially, and rigorously (ISSCR 2006).

The ISSCR guidelines also differ from the NAS guidelines in several important ways. First, while the ISSCR acknowledges the importance of commercialization, researchers are encouraged to make materials readily accessible to members of the research community. Institutions are urged to maintain nonexclusive access for the research community for the public's benefit, and not to have restrictive data and materials sharing agreements. A model uniform material transfer agreement form is provided (ISSCR 2006). Second, the two sets of guidelines agree that chimeras should never be permitted to mate with each other for ethical reasons. However, while the NAS opposes mating chimeras with non-chimeras, the ISSCR allows for the possibility after approval from a local review committee. In general, the NAS makes a distinction between the use of pluripotent cell lines (which require full SCRO review) and the use of differentiated derivatives (which may be permitted to have expedited review, depending on the case), but does not address teratoma assays. The ISSCR on the other hand distinguishes the creation of teratomas in animal models from the creation of chimeras for research. Teratoma assays are explicitly allowed under ISSCR guidance, since it has been shown that the cells forming a teratoma do not migrate to the central nervous system or the germline of the recipient animal (Lensch et al. 2007). In fact, arguments have been made that teratoma assays should be examined by an institution's animal care committee instead of undergoing review under the SCRO committees (Lensch et al. 2007; Daley et al. 2007). Third, the NAS does not address social justice issues, whereas the ISSCR does, asserting that the research and its benefits should yield worldwide benefits, instead of therapies for a chosen few.

Social justice issues are also considered by the Hinxton Group, another voluntary, interdisciplinary consortium on stem cell ethics and law. The consortium primarily aims to facilitate communication among researchers and policymakers, but has composed its own additional recommendations. Notably, the group advocates that scientists submit their derived stem cell lines to repositories, making them openly accessible, and seeks to engage the public in creating a research regime that balances scientific inquiry with social values. Like the ISSCR, the group advocates incentivizing data and materials sharing and creating licensing and patenting procedures that

[15] The ISSCR initially published its guidelines for stem cell research in 2006, available at http://www.isscr.org/docs/default-source/hesc-guidelines/isscrhescguidelines2006.pdf. Additionally, the organization published its Guidelines for Clinical Translation of Stem Cells in 2008, which deals with non-regulated clinical trials. See http://www.isscr.org/docs/default-source/clin-trans-guidelines/isscrglclinicaltrans.pdf.

"promote fair, reasonable, and nondiscriminatory (equitable) access to knowledge and health care applications" (The Hinxton Group 2006; Winickoff et al. 2009).[16]

Additional organizations and professional societies, such as the American Society for Reproductive Medicine (ASRM), International Society for Cell Therapy (ISCT), and the American Association of Blood Banks (AABB), have guidelines and recommendations for particular aspects of professional ethics, embryo acquisition and use, and laboratory practice relevant to their own constituencies and as such, they may emphasize certain aspects of good research practice over others (American Society of Blood Banks 2009; American Society for Reproductive Medicine 2009).[17] Such guidelines were established long before the existence of many contemporary cell-based technologies. For example, bone marrow transplantation has been practiced since the 1950s, and blood banking systems have been around since the 1930s (Lederer 2008). Both are based on clinically-oriented biological expertise and deal with the collection, processing, and storage of minimally manipulated materials. However, the guidelines did not anticipate interdisciplinary research practices, or the need to exchange materials across jurisdictional and policy boundaries. Furthermore, a recent court case has challenged traditional ways of handling autologous tissue, which will affect common procedures in regenerative medicine. An example is bone marrow-derived cells which are filtered, expanded, and possibly treated with agents to aid their proliferation, then re-injected into the patient. Previously, materials such as bone marrow or blood products removed and replaced in a clinic or hospital (and not entering into interstate commerce) have only had to follow good laboratory practices (GLP) but were not subject to full Food and Drug Administration (FDA) review. In *United States v Regenerative Sciences LLC*, the FDA won jurisdiction over such procedures when there is more than minimal manipulation or the cells are not used in a homologous manner. This requires additional review by the Center for Biologics Evaluation and Research and in some cases, clinical trials might be required. At the core of the dispute are definitions of "minimally manipulated" and "homologous use."[18] While not an ethical issue per se, the case demonstrates how taken-for-granted guidelines and practices may be challenged in the context of an emerging technology under public scrutiny, and

---

[16] http://www.hinxtongroup.org/wp.html.

[17] The ASRM Ethics Committee document on embryo donation can be found at http://www.asrm.org/publications/detail.aspx?id=285; the ISCT regulatory documents can be found at: http://www.celltherapysociety.org/index.php?page=regulatory; and the AABB regulation statement is located at: http://www.aabb.org/resources/bct/Documents/coi_ct1109.pdf. Organizations formed around engineering and informatics research have not yet addressed interdisciplinary, cell-based research, as they traditionally had completely different areas of focus for ethical practice, and have only recently become closely intertwined with biological research.

[18] Minimal manipulation is defined as processing that does not alter the relevant biological characteristics of cells or tissues (21 CFR 1271.3 (f) (2)). Homologous use is defined by the FDA as the repair, reconstruction, replacement, or supplementation of a recipient's cells or tissues with an HCT/P that performs the same basic function in the recipient as in the donor (21 CFR 1271.3.(c)).

raises questions about what regulatory authorities should be involved in research and clinical practice (Werner et al. 2012; McAllister et al. 2012).[19]

Other organizations, such as the Alliance for Regenerative Medicine and the Interstate Alliance for Stem Cell Research, track policy trends and promote the use of common standards and policies to facilitate coordination among state, organizational, and federal guidelines and regulations.[20] While some of these groups have international reach, others are regional- or country-specific.

In sum, there are a number of types of organizations at different levels that have attempted to provide guidance on various aspects of research integrity specific to stem cell research, primarily focusing on derivation and the treatment of embryo donors (primarily a human subjects protection issue). In the absence of coherent national and international agreements, such guidelines have allowed researchers to proceed with at least some consistency. However, some conflicts exist between such voluntary but authoritative guidelines and other national and international governance bodies, including the NIH (which exerts control through the control of funds) and regulatory bodies such as the FDA in the United States and the Human Tissue Authority and EU Council in the European Union, as well as other country-specific agencies (Levine 2011; Caulfield et al. 2009; Elster et al. 2008).[21]

6.3 Essential Components of Responsible Conduct of Research: Inside and Outside the Checkboxes

Having outlined the various administrative mechanisms involved, we turn now to specific areas considered to be central to the responsible conduct of research. Table 6.2 lists the RCR topics covered by the ORI and for which compliance is required by most institutions. The topics are applied to any area of formal research, including science, engineering, social science, and the humanities. Legislation, details about each topic, and resources can be found online at http://ori.hhs.gov.

[19] The state of Texas, however, approved rules allowing allogeneic and autologous adult-derived stem cells to be used experimentally in clinics after Governor Rick Perry received a stem cell treatment for back pain and became an advocate for adult-derived stem cell research (Cyranoski 2012).

[20] The Interstate Alliance on Stem Cell Research (IASCR) is a voluntary body whose stated mission is to advance stem cell research (human embryonic, adult, and other) by fostering effective interstate collaboration, by assisting states in developing research programs, and by promoting efficient and responsible use of public funds (Lomax and Stayn 2008).

[21] Numerous country-specific agencies deal with embryonic stem cell research, and in some countries more than one agency is involved. A few include the Canadian Institutes of Health Research, Danish Council of Ethics, Geschäftsstelle des Nationalen Ethikrat (German National Ethics Council), and Zentrale Ethik-Kommission für Stammzellenforschung (Office of Central Ethics Committee for Stem Cell Research in Germany, the Ministries of Science & Technology and Health in China, and the Ministry of Education, Culture, Sports, Science and Technology (MEXT)) in Japan, among others.

Table 6.2 Responsible conduct of research: essential components

RCR topic	Definition and requirements
Human subjects research	Human subjects must be provided appropriate protections according to the Common Rule, research protocols must be approved by a local Institutional Review Board (IRB); subjects must be informed about procedures and risks, and must be guaranteed confidentiality and the right to withdraw from the trial; the research should minimize potential harms; and subjects must be fairly selected with benefits and risks distributed fairly among participants
Publication/authorship	Publication of full, fair, and significant results by researchers who sufficiently contributed to the findings; data must be analyzed under reasonable statistical tests, all relevant data must be reported and be accurate; complete reference list must be provided
Research misconduct	Researchers must not intentionally or knowingly commit act(s) of plagiarism, fabrication, or falsification in any element of the research process
Animal welfare	Strive to reduce the number of animals used, refine protocols to minimize the animal's pain/distress, and replace the use of animals using other means when possible (for example, using cell lines)
Mentorship	Provide a relationship characterized by mutual respect and professionalism; mentor must take responsibility for mentee's RCR training
Data management	Ensure data integrity (no investigator bias, proper data collection and analysis techniques); researchers must permit sharing when warranted, have clearly defined ownership, retain data for a reasonable period (7 years recommended)
Collaborative science	Researchers should determine ownership of data and IP, have a clear and fair division of responsibilities, and determine authorship
Conflicts of interest and commitment	Research must be conducted objectively, uninfluenced by outside bias (personal, professional, financial); potential sources of other conflicts must be declared
Peer review	Manuscripts and materials should be reviewed competently, objectively, and without bias or conflict of interest

A few points need to be made about such lists. First, although these are the areas identified as the most relevant for RCR training, some may receive greater scrutiny and oversight than others, or may periodically receive extra attention. For example, the introduction of new research management arrangements has recently stimulated changes in policies on research subject protections and conflicts of interest, and the rapid expansion of data visualization tools has raised concerns about data misrepresentation and misconduct. We expand upon the discussion of human and animal subjects, misconduct, and COI for this reason. Second, the categories listed in Table 6.2 assume that research integrity can be tidily placed in discrete categories, and then dealt with procedurally. In the real world of everyday practice, dilemmas do not fit into such categories perfectly; they may fall across several categories or may be too ambiguous to properly classify. For example, a case of misrepresented findings by a senior researcher may affect mentored students or junior researchers involved in the project in addition to constituting misconduct. In other cases, there may be tensions between values underlying aspects of research integrity: for example, open sharing of data may conflict with ideas about respecting a biospecimen donor's autonomy.

A recent example demonstrates how multiple failures can occur across RCR categories or checkboxes, and how failures may ultimately endanger human subjects. A paper by Duke University researcher Anil Potti compared the molecular traits of various cancer tumors to determine which chemotherapy would be the most effective treatment. A team of biostatisticians were unable to duplicate the findings and found numerous errors, including mislabeled cell lines and transposed data spreadsheets, and tried to draw attention to the suspected misrepresentations (Baggerly and Coombes 2010). However, because scientific communities often do not read each other's literature and may not be looking for the same sources of error, the misconduct was not dealt with until other researchers in the medical field uncovered misrepresentations in Dr. Potti's resume, which led to an investigation of the veracity of his research findings (Couzin-Frankel 2010). This case of misconduct was particularly egregious because the paper was cited numerous times, and a number of clinical trials were initiated with patients randomized based on the markers identified by the Potti team. The trials had to be halted, and the risk to patients could have been significant. The Institute of Medicine report, "Evolution of Translational Omics: Lessons Learned and the Path Forward" (2012) uses the Potti case to highlight failures of the systems of integrity, including unclear lines of accountability, poor data management, a lack of independent confirmation of omics discoveries, a failure to solidify consistent test methods, a lack of validation of the omics-based test prior to beginning clinical trials, and various conflicts of interests that were undisclosed in informed consent documents. The failure of the system cannot be overemphasized: beyond the actions of a single individual, peer review, data sharing requirements, and reporting systems all failed until other individuals took it upon themselves to investigate. Importantly, the case also serves as an example of the kinds of problems that can be missed as research becomes increasingly interdisciplinary, with insufficient expertise in other fields to recognize problems as they occur.

The areas identified as essential RCR components, along with recommendations for practice and training materials, were written primarily by experts in ethics, policy, and the law. While they serve well as a general guide, in some cases, they may not be in tune with the many complex, interdisciplinary kinds of science that have emerged over the past decade or so. Many of the teaching illustrations used in training modules are more applicable to large-scale clinical research, drug discovery, or other areas which may be far removed from situations relevant for stem cell and regenerative science researchers. Furthermore, many of the modules have not taken into account the way novel tools and technologies raise new gray areas of ethical practice and challenge what constitutes "misconduct." Uncertainties about how to proceed are also raised in light of networked, large-scale, or international collaborations across labs, regulatory environments, and cultural contexts.

The complexity of techniques, tools, and the non-unified nature of guidelines puts considerable burden on researchers themselves to discern when a line of appropriate behavior has been crossed.

The next section looks deeper within the checkboxes of RCR components, providing illustrations with greater salience to researchers in cell and molecular-based biology and engineering and illustrates how contemporary research practices raise challenging questions for responsible research. We focus first on human subjects protections and scientific misconduct, because these areas are arguably the most critical to ethical scientific practice, and are fundamental to trust among researchers as well as between researchers, the public, and governmental or financial supporters of research. Aspects of human subjects protections for stem cell research are covered in the chapters by Resnik and King in this volume; however, we elaborate a few issues which are likely to continue to require particular attention both now and in the future. Additionally, proposed new guidelines for human subjects protections, should they be approved, will have a strong impact on research using human biological materials. We next examine ethical concerns arising from increased use of chimeras in research, which has not yet been sufficiently dealt with guidance on animal use. Then, as an extended illustration of potential misconduct, we use the case of visual data representation, as it is increasingly relevant to daily research and publication and presents a growing area of concern. The section ends with a discussion of COI, where issues reflect broader trends in contemporary research, but some challenges have arisen in stem cell-related research due to the particular nature of materials and financial and organizational agreements used. Recent changes in COI policy in the United States which will affect such arrangements are also noted.

6.3.1 *Human Subjects: Protecting Research Participants and Tissue Donors in Regenerative Research*

6.3.1.1 Oversight of Research Using Human Subjects

Until the twentieth century, experimentation on humans was done within the bounds of routine medical practice and left up to the physician's judgment. There was no formal guidance or consensus about what counted as appropriate experimentation, and little oversight other than professional codes. Eventually, the need to test the specificity of disease causation and the efficacy of vaccines and other medicines necessitated more systematic experimentation on humans. This exposed many more patients and healthy volunteers to possible harms. By the nineteenth century, some efforts to create protections for human subjects began; however, it was not until after major abuses became publicly visible that regulations were institutionalized. Beyond the well-known experiments conducted in concentration camps in Nazi era Germany, governments in the United States and other countries conducted numerous secret experiments using radiation exposure, toxins, psychological torture, and more, and clinicians also tested medical theories and treatments on humans, sometimes without their knowledge or permission (Annas and Grodin 1995; Marks 1997; Moreno 2001; Reverby 2009). Pharmaceutical and medical device companies also routinely tested products in vulnerable populations, often in countries where there was less likelihood of the work being made public (Petryna 2007). Objection to such practices culminated in the Declaration of Helsinki (1975), a consensus document by the World Medical Association and now considered to be the foundation of modern research ethics. In the US, the National Commission for the Protection of Human Subjects of Biomedical and Behavioral Research was created, and its subsequent publication of the Belmont Report in 1979 put forward basic principles of research-specific protections for human subjects (Levine 1988).[22]

Today, any time humans are used in formal research, as subjects in a trial (such as a test of therapeutic devices, drugs, or other substances), as a source of biological material (blood, cells), or when information from humans is used (data from the genome, microbiome, identifiable medical histories, etc.), strict guidelines exist to ensure the subjects' safety, confidentiality, and voluntariness.

All work that is supported or conducted by the U.S. DHHS in the United States is subject to its human subjects protection regulations, codified at 45 CFR part 46, subparts A through D. In 1991, 14 US federal agencies agreed to implement uniform regulations for the protection of human subjects as a part of the code of federal regulations governing health and human services. The resulting regulation is known as the "Common Rule" and serves as the primary guidance for all federal agencies in the United States. Similar principles are used in other countries.[23] Clinical research subject to FDA review (that is, any product used in or on the human body and intended for interstate trade) is covered under 21 CFR 50 and 56 (human subjects protections relating to informed consent).

Two trends in research practice have raised questions for RCR. While academic and federally funded research institutions have structures in place to oversee compliance, for-profit product sponsors have moved away from using such centers to conduct clinical trials. Increasingly, they use contract research organizations and for-profit IRBs to review protocols and human subject recruitment and protection plans,

---

[22] The principles are beneficence (commonly parsed as "do no harm"), respect for persons (individuals should have a right to autonomy), and justice (subjects should not be exploited and procedures in research should be fair and reasonable). These remain the cornerstone of human subjects protection regulations. The Belmont Report can be found at http://science.education.nih.gov/supplements/nih9/bioethics/guide/teacher/Mod5_Belmont.pdf). Further reading on these principles, including critiques, can be found in Jonsen (1998) and Levine (1988), among others.

[23] The National Research Act of 1974 required all institutions involved in federally funded research to create Institutional Review Boards (IRBs), committees at each institution which review, approve, and monitor research involving human subjects. The following agencies and departments have signed onto the Common Rule: Agency for International Development, Consumer Product Safety Commission, Department of Agriculture, Department of Commerce, Department of Defense, Department of Education, Department of Energy, Department of Housing and Urban Development, Department of Justice, Department of Veterans Affairs, Department of Transportation, Environmental Protection Agency, National Aeronautics and Space Administration, and the National Science Foundation.

which has raised concerns about the objectivity and ethics of for-profit research practices (Fisher 2005).

Another increasingly frequent practice is the use of existing samples or databases collected for another purpose in a new study, without seeking consent from the original donors. It is important to note that the "human subject" label includes any specimen or data derived from humans for which any individual personal information (medical history, identifiers such as record number, address, etc.) can be readily identified. In such cases, consent from donors must be obtained prior to use in research. Specimens that have been anonymized through the use of codes or stripping of all identifiers are not counted as human subjects.[24] In the United States, the privacy of personal medical information is additionally protected under the Health Insurance Portability and Accountability Act of 1996 (HIPAA), under the auspices of the Office of Civil Rights in the DHHS. There are no restrictions on the use or disclosure of de-identified health information under HIPAA. However, should it be necessary to contact donors, for example, to gain additional information from their medical records to analyze along with their biological specimen, conflicts may arise between the different laws covering individuals' privacy.

Existing RCR training material deals primarily with the protection of human subjects participating in conventional clinical research trials. However, the constantly changing landscape of medical research has raised new questions about how best to protect human subjects, both in clinical research and research using information from human specimens.

6.3.1.2 Human Subjects and Stem Cell Research

For regenerative medicine research, the more visible areas of concern have to do with the fact that most of the techniques have not yet been tried in humans, and, therefore, may have unknown effects not observed in animal or laboratory studies (see King, Chap. 9, this volume). This is even more of a concern with the rapid increase in the number of trials and advertised therapies appearing worldwide which may have poor transparency regarding the protocols and types of materials used. Ethical issues regarding individuals who participate in such therapies raise difficult questions about patients' exposure to potentially highly risky procedures, and sometimes the limits to scientific freedom are pitted against the enforced regulation and oversight by national or international bodies. The problem of "medical tourism," that is, patients seeking clinics which offer innovative therapies but which may not abide by generally accepted oversight rules, or travelling to clinics in countries with

[24] Personally identified information is that which can be directly tied to an individual, including name, geographic information smaller than a state, social security number, birth and death dates, phone and fax numbers, email and IP addresses, medical record and health plan numbers, vehicle license numbers, device serial numbers, and biometric identifiers (including voice print and photos) (45 C.F.R. § 160.103). A summary of the rule can be found at http://www.hhs.gov/ocr/privacy/hipaa/understanding/summary/index.html.

less regulation or ethical oversight, has become highly visible in recent years because of several deaths and significant injuries (Gunter et al 2010; Hyun et al. 2008; ISSCR 2008).

Where human subjects protections for regenerative science research differ somewhat from other areas is in the use of biological materials from humans. A variety of human biological materials are used for many of kinds of research. However, the use of human embryos is particularly sensitive. In many countries, questions of whether or not it should be morally permissible to create embryos from donated gametes for the purposes of research, or to create embryos that could produce children (somatic cell nuclear transfer or reproductive cloning), and the protection of the embryo donor's privacy have become battleground issues in the context of political and social conflict about the use of human embryos.

People who donate their bodily materials (blood, gametes, tissue) for research or therapy may do so as an act of altruism or in support of science, or because they have an interest in a particular type of research. Individuals affected by a disease may be approached to contribute a specimen to research that disease or to help develop a therapy. Tissues may also be acquired as a byproduct of individuals having a procedure, for example, when tissue is removed in surgery, or in the case of hESC research, when a couple is creating embryos for the purpose of having children, but not all the created embryos are used. As Resnik discusses in Chap. 7, there are general protocols for dealing with donations of biospecimens: donors are promised anonymity and must give informed consent for their specimens to be used in research. While some of the issues are the same as using say, blood or tumor cells for testing, immortalized cell lines may create new challenges due to the ability to use them over long periods of time, and to use the lines (and data produced by them) for purposes far beyond what the donors had imagined in their consent. There are unresolved questions about how best to ask donors for consent, when there may be many reasons to use a line for additional research beyond what was mentioned in the consent form. It is not always known which lines may have the best clinical utility, and new experiments may become possible that did not exist at the time of specimen collection. Still, it may be almost impossible for donors to withdraw if they reconsider their consent (Lowenthal et al. 2012).

For this reason, some observers have advocated for language in the consent form in which donors agree to a broad variety of possible procedures ("broad" consent), while others argue that this diminishes autonomy, and donors should be given the opportunity to consent to very specific procedures which exist at the time of consent (Aalto-Setälä et al. 2009; Hansson et al. 2006; Scott et al. 2012). A "sustained interaction" system, in which traceability is maintained and recontact is allowed, either for additional permissions, to report incidental findings or the results of a study, or for other circumstances has been proposed to allow human subjects flexibility (Isasi et al. 2011; Lowenthal et al. 2012). Coded identifiers attached to specimens would be maintained, rather than de-linking identifiable information as is currently done. This system would allow some categories of research to go forward with only the initial consent, but others (mainly sensitive research areas such as the production of gamete cells) would potentially require recontact and additional consent.

A primary advantage, as the authors see it, is the development of trust and a sense of collaboration with researchers, and greater transparency regarding secondary research while allowing donors to retain some control over the disposition of their specimens. The disadvantage is the necessity to create and maintain a new infrastructure to track and monitor individual choices at various points over time, which researchers are likely to see as burdensome and costly.

Another key problem is that of protecting biospecimen or embryo donor anonymity. There may be conflicts between the desire to protect anonymity and the need to know information about donors, both to be able to contact them regarding their original consent and to use the information to conduct certain kinds of research, such as disease-specific studies. Donors of materials for hESC or iPSC research trust that the donation will be anonymous (e.g., their personal information will not be connected to their biological sample), that researchers will not share information with others without their permission, and that the material will be used for the express purpose for which they consented. In fact, oversight rules require that no personal information should be collected at all unless it is essential to the protocol for which the donor is consenting to provide tissue. This is particularly important when the tissue—or particular uses of it—is sensitive, as in the donation of embryos, fetal tissue, or gametes. When that trust is broken, the entire system of tissue collection for research and therapy may suffer.

An issue of growing concern is data handling and privacy; for example, when data produced by biological materials and collected on the basis of one set of consent agreements is used for a different purpose than one to which donors were recruited. As genome sequencing has advanced, it has become far easier to identify biological materials and match them with donors, making a guarantee of protecting donor anonymity virtually impossible (see Resnik Chap. 7 and Ossorio Chap. 5, this volume). Gymrek et al. (2013) easily identified individuals in a study group, as well as their relatives not involved in the study, using simple Google searchers and basic, openly available information published from the 1,000 Genomes Project (see also McGuire et al. 2008).

Elaborate systems have been invented to "de-identify" samples to maintain anonymity (Elger and Caplan 2006; Resnik, Chap. 7, this volume). Ossorio points out that in the era of large genetic databases, it becomes almost impossible to guarantee anonymity, even though informed consent forms promise such protections (Ossorio, Chap. 5, this volume). Yet there are other breaches of anonymity arising from the very different priorities and needs of researchers, federal agencies, clinics, and regulators, and in the case of stem cell research, the recent federal guidelines for hESC research.

To illustrate, the 2009 NIH guidelines (which affect federally funded US researchers) require that only cells from approved ES cell lines be used, and in order to be approved, there must be proof that proper informed consent was obtained from embryo donors. While this was straightforward for new lines going forward, the numerous lines which had been in use globally for many years were not grandfathered into the new guide. This meant either that those institutions responsible for the lines had to retrace original donor consent processes (and if found not to be in compliance with the new guidelines, the lines could not be used), or that donors might have to be re-identified, located, and re-consented under the new standards. Since many of these lines were created nearly 20 years earlier, locating donor couples made the re-consenting process onerous.[25] Also, if original donors were consented with the intent of using material for one particular type of research, such as diabetes, according to the guidelines, researchers could not use the line derived from that embryo for any other purpose, such as cardiovascular research, even if the cells proved to work well for the other purpose (Taylor 2009).

An analysis of consent forms showed considerable inconsistency in what was explained to donors among the various clinics, institutions, and for-profit organizations which had derived earlier lines (Streiffer 2008). This is not surprising, since informed consent has never been a standardized process across types of research or organizations. The new guidelines impose a strict separation between the act of undergoing IVF treatment and the donation of unneeded embryos, to ensure voluntariness. Because couples may plan their families over several years, the request for donation may come many years after the embryos were initially frozen and stored. If donated, further information about medical history may be required of the donor couple for some studies. At the same time, the FDA, concerned with the transmission of infectious agents into potential recipients, began requiring information from embryo and gamete donors from which cell lines have been derived at the time of the IVF procedure.[26] This in essence contradicts the NAS guidelines' recommendation to separate donation from IVF treatment, and may necessitate breaking anonymity to gather information from couples who may not have considered donation at the time of their original treatment (Hogle 2011).

Another problem exists when cell lines are exchanged among researchers (Lo et al. 2009). Institutions share their cell lines and other biological materials with other researchers and institutions using a formal binding legal contract, called a Materials Transfer Agreement (MTA). When the material being transferred is subject to specific restrictions (for example, research purposes limited by the donor's consent), the donating institution will insert clauses permitting the receiving institution to use the material for the specified purpose only (and prohibiting the use of the material for unauthorized use). However, there is little ability to audit actual uses to ensure that they comply with the agreement and stick to the uses for which donors gave consent.

---

[25] Significantly, considerable data had already been published from several lines, a few of which have become the "gold standard" for research. If those lines had been disallowed due to inability to trace the donor consent process, the result could be catastrophic, raising questions about how far a rule intended to protect donor privacy should go without damaging the research enterprise.

[26] 21 CFR 10.115(g)(4)(i) Part 1271 subpart D, effective May 25, 2005 now applies to donors of cells and tissues including hematopoietic stem cells derived from peripheral and umbilical cord blood, reproductive cells and tissues, in addition to tissues previously covered by statutes (part 1270), such as human dura mater, human heart valves, and other tissue for transplantation. Part 1271 is far broader than the prior 1270 which was primarily for biological entities, because it applies to HCT/Ps regulated in any category of regulatory entity (e.g., drug, device, or biological). Exempted are the Wisconsin lines, which were created before 2005.

Finally, there may be a question of whether the materials are acquired either without a person's knowledge or with coercion.[27] A well-known illustration in regenerative biology is the case of South Korean researcher Hwang Woo-Suk, who claimed that he had successfully cloned human embryos. There were multiple problems with the research, the most spectacular being the falsification of data.[28] A significant ethical breach, however, involved alleged coercion of his female laboratory workers into donating oocytes for the research. Inducing individuals to participate in research through offers of money or services, or as in this case, exploiting the power relationship of a senior researcher with his subordinate laboratory staff, violates human subjects protections by most standards. Particularly because this case arose in the highly sensitive and contested area of human cloning, the practice was especially egregious. Not only did Hwang's actions destroy the hopes for a successful Korean research enterprise (at least initially), but the entire affair cast a pall on somatic cell nuclear transfer research as a whole (Gottweis and Triendl 2006; Normile et al. 2006; Snyder and Loring 2006).[29]

6.3.1.3 Proposed Changes to Human Subjects Protections

Recently proposed changes regarding human subjects protections, listed in Table 6.3, will directly affect regenerative science researchers should they be approved. The most significant changes reflect concerns about donor privacy and the identification of human subjects, as described above. The rapid growth in shared databases and biorepository information has raised a host of new legal and ethical issues. One change being considered is whether donors should be contacted for each new use of data from their donated materials, or whether research using preexisting data and biospecimens, tests, and surveys should be excused from review by an IRB.

[27] Coercion exists if individuals are offered incentives that may induce them to participate when it may or may not be in their best interest, or when they feel the implications of not participating.

[28] The case is well known and will not be repeated in detail here, but see, for example, Gottweis and Triendl (2006), Beasley et al. (2002), and Lee and Park (2006). The Hwang Woo-Suk case further illustrates the international variation in definitions of misconduct, investigatory procedures, and penalties. Hwang was convicted of bioethics violations and embezzlement in his native South Korea, sentenced to 2 years in prison (suspended) and his funding was withdrawn, but he was not convicted of fraud. A senior coauthor, Gerald Schatten (University of Pittsburgh), was found to have committed "research misbehavior" (an ill-defined, less severe version of research misconduct) for accepting senior author status while not verifying the accuracy of the data and participating significantly in any of the experiments (Marris and Check 2006). He was not suspended from continuing research.

[29] The question came up again when the Human Fertility and Embryology Authority in 2007 authorized the payment of women for oocyte donation to support stem cell research. Two long-standing questions were raised: whether offering money in exchange for bodily materials unduly places women at risk for harm (whether or not they willingly donate) and whether payment for human reproductive materials which may be used to create an embryo is ethical in any case (Baylis and McLeod 2007). While it is beyond the scope of this chapter, information about payment for gamete and embryo donors can be found in the NAS guidelines at 82–89.

Table 6.3 Human subjects protections—proposed changes

Area	Current rule	Changes being considered
Data security	No current specific data security protections for IRB-reviewed research	Specified data security protections would apply to the level of identifiability of the collected information
Biospecimens	Research using existing biospecimens (clinical or from prior research) can be done without consent by stripping the specimens of identifiers	Written standard consent required for research use of all biospecimens regardless of identifiability; broad consent permitted if specified
Informed consent	Current provisions of the Common Rule provide only basic information about the elements of informed consent and how consent documents should be written	Revise regulations to provide greater specificity and clarity about how consent forms should be written and what information they should contain, with goal of aiding individuals to make good decision about participation in studies
Studies subject to common rule	Federal protections only apply to studies that are funded by certain federal agencies (Common Rule agencies), or to clinical investigations that involve products regulated by the FDA	Regulations would apply to all studies involving human subjects conducted by US institutions which receive federal funding for research from a Common Rule agency, regardless of funding source
Adverse events	Adverse events and unanticipated problems occurring in research are reported to multiple agencies and with various timelines, with no central database as a repository for such data	A single website meeting federal reporting requirements would be created for the electronic reporting of all such events, leading to a single database and harmonizing reporting requirements across agencies
IRB review	Each site in a study requires IRB review. Although one IRB is permitted to review multiple sites, it is common for a single study conducted at multiple sites to have many IRBs separately conducting reviews	For all of the US sites in a multi-site study, a single IRB of record is proposed
Common rule agencies	Common rule agencies and the FDA are authorized to issue separate guidances for human subjects protections (lack of uniformity between agencies)	If agency to agency differences are not justified by differences in the applicable statutes or missions, guidance will be made more uniform
Exempt studies	Six categories of studies qualify as "exempt" from the regulations, meaning that they do not have to comply with any of the regulatory requirements	Currently exempt studies would no longer be fully exempt from the regulations and would be subject to new data security protections; studies using biospecimens would have new consent requirements

(continued)

Table 6.3 (continued)

Area	Current rule	Changes being considered
Use of existing data and specimens	Permitted for research if sources publicly available or if information recorded by researchers in way that subjects cannot be identified, directly or through identifiers linked to them	Eliminate requirements that all data/specimens must exist at commencement of study and that researchers cannot record and retain subject identification information; to obtain and record such information, the subject's consent would generally be needed (but could be obtained when sample was collected with open-ended consent to future research uses); with regard to studies using existing biospecimens, see Biospecimens, *supra*

Currently, if materials are de-identified, no further permission is required, but evidence of the ease with which materials can be identified has challenged the belief that de-identification is sufficient. There is considerable disagreement about how to proceed, and differences exist between the views of bioinformatics experts, genomics researchers, disease advocacy groups, and privacy advocates.

Additional proposed rules would expand Common Rule coverage to all studies conducted at institutions receiving money from any of the 15 federal agencies using the Common Rule. This means that even studies sponsored by a for-profit company and conducted at a university receiving federal funds would be covered. For multi-institutional clinical trials, a single IRB is proposed to oversee studies, in lieu of current practices of IRBs at each location conducting oversight. This is intended to harmonize and speed the process of oversight.

At the time of publication, no action had been taken on the proposed rules. Whether or not they are finalized in the form in Table 6.3 changes will certainly be made regarding biorepository data that will affect regenerative medicine. Regenerative medicine also uses animal subjects extensively; therefore, the next section talks about oversight of research using animals.

6.3.2 Oversight of Research Using Animals, Including Chimeras

6.3.2.1 Animal Use and Welfare: Background and Regulations

The welfare of animals was a concern of scientists long before the protections of humans in research. Anti-vivisection and other animal care movements, particularly in the nineteenth century, led to animal protection legislation in Britain in 1876. By 1908, the American Medical Association had created guidelines for the care of laboratory animals but the first US laboratory animal protection laws were not passed until 1966 and only in 1979 were protections institutionalized nationally (Lederer 1997). Protections are based on three principles posed by Russell and Burch in 1959. Specifically, animals in experimental systems should be replaced with cell culture systems or lower order organisms when possible, experimental design and procedures should be altered to reduce the pain and suffering of animals used, and the number of animals used in an experiment should be reduced when possible.

Today, animal welfare is arguably one of the most thoroughly regulated areas in research, both on the federal and local levels. On the national level in the United States, animals must be kept and cared for in compliance with the Animal Welfare Act (7 U.S.C. §2131 et. seq.) and nine federal principles adopted in 1985 by the Office of Science and Technology Policy.[30] Additionally, NIH policy 6380-2/54206, "Responsibility for Care and Use of Animals" (2004) followed the Public Health Service's 2002 Policy on Humane Care and Use of Laboratory Animals.[31] The policy delineates standards of care for housing, pain management, and sacrificing of vertebrate animals used for research and testing. The National Academies of Science also publishes a guide for the care and use of laboratory animals (NAS 1996). The use of laboratory animals in federally funded research must also be overseen by the home institution's Institutional Animal Care and Use Committee (IACUC). The IACUC reviews research protocols and evaluates the institution's animal care and use, reporting annually to the NIH's Office of Laboratory Animal Welfare (OLAW). The OLAW requests that institutional IACUC committees consider the "3R" principles (replace, refine, and reduce) when evaluating grant applications and reviewing proposals. The Animals (Scientific Procedures) Act of 1986 provides similar regulation in the UK.[32]

With the increasing use of transgenic animals and chimeras, new questions have arisen regarding not only the nature of animal suffering but also the ethics of genetically combining species. Chimeras, defined as organisms containing cells or tissues combined from different animals, are increasingly common in medical research, and can be extremely valuable research tools.[33] Chimeras may provide valuable information about tissue function or disease causation where simulations are insufficient and testing in living humans would be unethical. There are several ways scientists are currently mixing species genetic material: by genetically altering an animal to make a human-like protein or to manifest more human-like functions in vivo, by populating a test animal (usually rodent) with human cells to test a drug or recapitulate a human disease, or by introducing early-stage human cells into a non-human embryo, for example, to study development. An important use in stem cell research is the use of teratoma assays to check the pluripotency of cell lines. Test

[30] The nine principles can be accessed here: http://grants.nih.gov/grants/olaw/references/phspol.htm#USGovPrinciples.

[31] The PHS document can be found here: http://grants.nih.gov/grants/olaw/references/phspol.htm.

[32] http://ec.europa.eu/environment/chemicals/lab_animals/legislation_en.htm.

[33] These can be mutated cells of the host organism or cells from a different organism or species (Human Genome Initiative glossary, found at http://www.ornl.gov/sci/techresources/Human_Genome/glossary).

cells are injected into a mouse followed by a growth period, and if they develop into a teratoma containing cells from all three germ layers, the test cells are proven to be pluripotent (Lensch et al. 2007). Other possible uses of human–animal chimeras involve replacing only a select population in a fetal animal with human cells differentiated in vitro. These may result in the development of mice with human neurons or with human cardiac cell implants (Laflamme et al. 2007; Muotri et al. 2005).

Human DNA may also be fused with enucleated animal eggs to form a "cybrid," that is, a hybrid cell which combines the nuclear genome from one source with the mitochondrial genome from another source. This is not a chimera, since it would have almost entirely human DNA, and thus has been put forward as an alternative to making embryonic stem cell lines requiring donation of human embryos. Cybrids could also be used to study the contribution of mitochondrial DNA (Ishikawa et al. 2008; Knowles 2003). The Human Fertilization and Embryology Authority (HFEA) specifically considered this case, and after consultation with public groups and scientific experts, decided to permit the use of cybrids for research purposes. There were considerable criticisms of the research in the UK, and questions about the legal status of such entities (Baylis 2008). Bahadur et al. (2008) questioned whether they would qualify as human embryos, and if not, perhaps they would not come under HFEA jurisdiction.

Many people are troubled by the moral, political, and legal concerns about incorporating human DNA into other species. In particular, there are concerns about human cells integrating into the nonhuman host in ways that might make the animal "too human-like" in appearance or behavior (Bobrow 2011; Karpowicz et al. 2005; Greely et al. 2007). The nature and extent of moral repugnance has undergone considerable debate, with no consensus about what constitutes a moral violation, except that most people agree that integration into the neural system and reproductive tissue is unacceptable.[34] Issues might also arise if human cells were introduced into animals later than the embryonic stage of development, since migration to other tissues might be unpredictable. Introducing cells—particularly human neural cells—at an early stage in nonhuman primates might also be a concern, as there may be greater potential to affect an animal's cognitive abilities earlier in the animal's development, particularly when the species is genetically and morphologically similar to humans. Some institutions place limits on development stage and disposition of chimeras as a result.[35] For some, it is important that donors be informed whether

[34] Arguments center around the ambiguous moral status of creatures that are neither fully human nor animal (Robert and Baylis 2003). Critics of this concept argue that the focus on species identity is based upon an incorrect view of human personhood (Haber and Benham 2012; Siegel 2003). Some ethicists conclude that "status-enhancing" research (such as integrating human stem cells into animal brains) should not be conducted unless it has minimal risk and probable therapeutic benefits to the subject (Streiffer 2005). However, it is not clear that chimera creation is, in fact, status-enhancing (id.).

[35] At the University of Wisconsin-Madison, any experiment involving the introduction of hESCs into embryonic animals past Carnegie Stage 23 (which is E16 in mice) is prohibited. The UW policy on chimera use in stem cells can be found at: http://www.grad.wisc.edu/admin/committees/scro/documents/ChimeraGuidanceDocument090527.doc.

cells from their embryo would be mixed with an intact embryo, either human or nonhuman, and a few existing ESC lines (for example, WA01, WA07, WA09, WA13, and WA14) were created with the express guarantee to donors that they will not be mixed (Streiffer 2005). Ethicists disagree on the nature and seriousness of species mixtures. One possibility might be to allow experiments to be designed in such a way that viable offspring of chimeras would not result (for example, putting reproductive cells in a body location where they could not be fertilized), or a method could be devised to prevent development beyond a certain stage.

Most countries cover the generation and use of animal models under animal use and welfare regulations, but do not have specific guidance for the growing number of applications for chimeras. The exceptions are the NAS (in the United States) and the ISSCR. In the UK, a working group of the Academy of Medical Sciences recommended the creation of additional oversight for some experiments, including introducing ESCs into nonhuman neural systems (Academy of Medical Sciences 2011). In Germany, the debate has a somewhat different flavor, since preservation of the dignity of humans is the core principle for post-World War II Grundgesetz (Basic Law). The German Ethics Council determined that mice carrying human genes are ethically acceptable, but other uses should require permission from a national ethics panel (for example, making transgenic nonhuman primates). On the other hand, introducing animal material into the human germline, experiments that would lead to the development of human sperm or eggs in an animal, and implanting an animal embryo into a human should be forbidden (Deutscher Ethikrat 2011).

Beyond the protection of human and animal subjects in research, researchers also have to produce reliable reproducible data for the progress of the wider scientific community. The next section on research misconduct provides some examples of recent transgressions and other possible transgressions in research conduct.

6.3.3 *Research Misconduct*

Highly visible disclosures of scientific fraud and misrepresentations in the 1980s brought scientific misconduct into public attention.[36] The Office of Science and Technology Policy in the White House ultimately created the Federal Research Misconduct Policy in 2000.[37] All federal agencies or departments supporting intramural or extramural research are now required to implement the policy through

[36] In an often-cited case, William Summerlin claimed to have transplanted donor tissue into a genetically unrelated recipient while avoiding graft versus host rejection. As proof, he presented white mice with spots of black fur he claimed came from donor mice. In reality, he had colored spots on the white mice using felt-tipped pens, which was discovered by one of his technicians (LaFollette 2000). In another case in the early 1980s, John Darsee, a promising young researcher of Brigham and Women's Hospital, was found to have fabricated data throughout his career, leading to the retraction of eight papers and 45 abstracts. Congress became involved when delays in the investigation and proceedings led to allegations of press cover up (id).

[37] Available at: http://ori.hhs.gov/federal-research-misconduct-policy.

their own policies or regulations.[38] Early on, NSF and PHS policies encompassed broad definitions of what counted as misconduct. However, the ambiguous nature of varied suspect acts of misconduct and the degree to which some infractions may be seen as serious made determinations of misconduct problematic. (Is adding a non-participating established author to a publication misconduct, or just doing something to increase its chances of publication?) After considerable debate about the proper scope of federal oversight, a new definition was written in 2005 (United States Department of Health and Human Services 2005). The new definition of research misconduct is the "fabrication, falsification, or plagiarism in proposing, performing, or reviewing research, or in reporting research results."[39] The federal rules for misconduct were intended to accomplish three things: clearly define misconduct while creating a consistent standard of proof, establish procedures for reporting and investigation potential cases, and provide protection for whistleblowers. Misconduct, as defined above, occurs when a researcher's act significantly departs from accepted practices, or when it is done intentionally or recklessly. It must be proven by a preponderance of the evidence.

The scope of the ORI's responsibilities regarding misconduct was modified in 2000 to become more focused on education and training, reflecting the prevailing sentiment that misconduct should be a matter of institutional and individual vigilance, and that government's role should be reduced. While this may provide more flexibility and decentralization, it also means that there may be variances among institutional policies. For example, some universities or research institutions may generally adopt the 2000 federal policy, others add or modify provisions, while still others may adhere to the earlier, more stringent policies set forth by the NSF and PHS. Researchers should thus know their own institution's policies as well as those of any institution with which they will collaborate.

The European Union defines misconduct somewhat differently, focusing instead on outcomes. It is defined as research involving human, animal and plant pathogens, toxic chemicals, and radioactive material that when misused could cause severe harm to humans, animals, plants, or the environment (European Research Council 2012). In addition to criminal actions (including fraud, falsification, and plagiarism), EU policy includes discussion about actions by terrorists or unethical military purposes, as well as research which can result in stigmatization and discrimination against populations and has considered the application of specific kinds of technologies, such as surveillance technologies and data mining or data profiling activities.

The difficulty, of course, is that both the US and EU policies, as written, leave considerable gray areas. Many cases may be a matter of interpretation, difference of

---

[38] The Department of Health and Human Services, Department of Defense, Department of Labor, Department of Transportation, Department of Veteran Affairs, the Environmental Protection Agency, National Aeronautics and Space Administration, National Endowment for the Humanities, the National Science Foundation, and the Smithsonian Institution have done so, and have published their policies online. The Department of Energy has published a Notice of Proposed Rulemaking. At the time of this writing, the Departments of Agriculture, Commerce, Education, Interior, and Justice are still in the process of drafting and reviewing policies.

[39] Federal Register, Vol. 70, p. 28370, May 17, 2005. Codified at 42 CFR Part 93.

opinion, or simply human error. Furthermore, the burden of judgment about such instances rests on local level institutional review authorities and ultimately, on the researchers themselves, to discern what constitutes misconduct. It becomes even more sensitive when a junior researcher or student observes dubious activities of a senior researcher or author on a grant or paper or in situations of whistleblowing.

Serious misconduct, such as deliberately falsifying data, is thought to be rare. Some have argued, however, that cases are significantly underreported, possibly because of the ambiguous nature of what counts as misconduct, or because researchers do not know reporting procedures, or are concerned about retaliation (Martinson et al. 2005). A surge in the number of retractions in scientific publications has been reported. For example, the number of articles submitted to Thomson Reuters Web of Science has increased by 44 % in the past 10 years, but the number of retractions has increased almost tenfold. About 44 % of these were attributed to acts of misconduct and 28 % were attributed to honest error (Corbyn 2012; Van Noorden 2011). A more recent study was even more damning. A survey of 2,037 retracted biomedical and life-science journal articles through May 3, 2012 found that 67.4 % resulted from misconduct. Fraud or suspected fraud was the cause of 43.4 % of the retractions. Duplicative publications were 14.2 % of retractions, plagiarism was 9.8 %, and only 21.3 % were due to innocent error (Fang et al. 2012). In a separate study, almost 2 % of surveyed scientists admitted to engaging in serious misconduct and 33.7 % admitted to questionable practices (Fanelli 2010).[40]

We illustrate the difficulties of such gray areas with an extended discussion of one area of increasing concern; that is, the representation of data, particularly visual images, in publications and grant proposals. As described below, there can be considerable blurriness between misconduct and simple misjudgment or human error.

6.3.3.1 Illustrating Gray Areas of Misconduct: Image Manipulation

Increasingly, such questions of research integrity in publications, grants, and other venues involve the mishandling of visual representation of data. One report showed that in 2008–2009, 68 % of all scientific misconduct cases reviewed involved the manipulation of images (Krueger 2009). Visual displays of data are the primary way of communicating findings, and act as evidence supporting experimental results. With the widespread adoption of digital image processing software, however, questions have been raised about what constitutes inappropriate manipulation of images. There may be a blurry line between enhancing images (to make them clearer, to highlight features of a finding, or to diminish artifacts) and manipulating images in a way that could alter an interpretation or misrepresent findings. Software allows easy cutting and pasting or rearranging of images, but other kinds of modifications can also be made, including adjusting settings while taking a photo, intensifying or changing a color, or highlighting or diminishing immunofluorescent markers.

[40] Interestingly, those surveyed claimed that 14.2 % of their colleagues had engaged in serious misconduct and an overwhelming 72 % in questionable practices, either indicating a lack of trust in colleagues, or, possibly, observation of underreported incidences of misconduct practices.

The most frequent problems come from manipulating images from immunolabeled blots and gel electrophoresis and photographs from various microscopy technologies. As a result, key journals have created guidelines for their own publications, and some are using forensic experts to detect when images have been changed inappropriately (Frow 2012).[41] The ORI has also created Photoshop "forensic droplets" that aid its reviewers in finding image manipulation in submitted figures.[42]

Most journals are limited to independently investigating only the most incredulous claims due to staffing and time constraints. The *Journal of Cell Biology* has a full-time employee dedicated to checking the figures of each accepted paper for veracity, and estimates that nearly 50 % of the papers require a figure to be reworked because it does not meet the journal's standards. Approximately 1 % of the papers accepted contain figures where the "efforts to manipulate are so egregious" the journal revokes its acceptance (Maher 2012).

A few highly visible cases of misconduct involving image reproduction have drawn attention to the problem, particularly in molecular biology and related areas. Perhaps most notorious was a case involving stem cell research. Phase-contrast photos submitted by Hwang Woo-Suk were used to claim that 11 embryonic stem cell lines had been successfully created using somatic cell nuclear transfer technique (cloning). Investigation showed that they were actually overlapping images of one cell colony, and that the images offered as negative controls were also manipulated (Saunders and Savulescu 2008).

Misrepresentation may or may not be intentional: in fact, it may be difficult for some researchers, particularly new learners of a particular laboratory technique or imaging software, to discern when alterations to an image may constitute misconduct. However, regardless of the researcher's intention in image alteration, the result could be retraction of a publication or withdrawal of a grant. Composite images are often required for making scientific arguments, but readers need to be informed how the composites were generated and readers must have access to the raw images.

There are many gray areas within image manipulation. For example, two Israeli researchers published an *International Immunology* article in which the Western blot lanes were derived from multiple experiments and made into a composite (Safadi and Pappo 2007). The authors did not disclose that the image was a

[41] Some journals have specific standards regarding image submission. For example, *Nature* acknowledges that "a certain degree of image processing is acceptable for publication (and for some experiments, fields and techniques is unavoidable), but the final image must correctly represent the original data and conform to community standards," but has guidelines for submission, including retention of the original, untouched image. The *Journal of Cell Biology* screens all figures for any evidence of manipulation, and sets forth examples of such manipulation; for example, "no specific feature within an image maybe enhanced, obscured, moved, removed, or introduced."

[42] "Forensic droplets" detect image alteration by searching for similarities and differences between images. The program color-codes each pixel in the images and superimposes them onto a single image for comparison. Common features appear in red, while unique features remain black or white, helping editors pinpoint potential erasures. Editors also use contrast enhancement and histogram equalization which reveal weak borders around objects (due to a cut-and-paste manipulation), or areas where an object has been removed from an image.

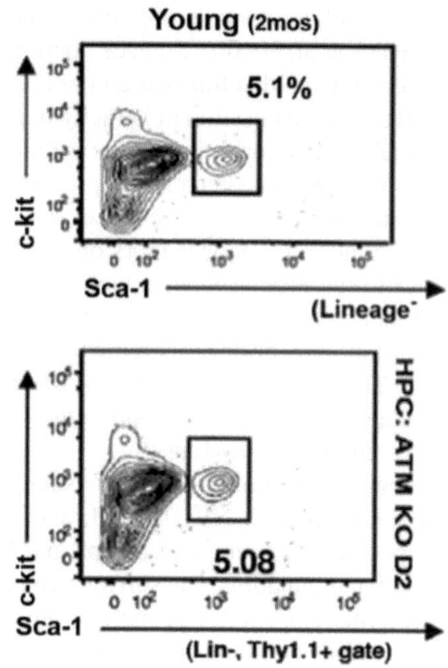

Fig. 6.1 The plot on top is figure S 3b from the 2010 *Nature* article while the seemingly identical plot below is Fig. 6 from the 2008 *Blood* article (Katnelson 2010)

composite, and later recommended the article's retraction (Safadi and Pappo 2011). While the researchers did not attempt to hide the nature of the blot, the lack of explicit disclosure could lead less-experienced researchers astray.

Another example involves flow cytometry, a commonly used laboratory technique to quantify cell populations by measuring various cell components through the use of fluorophores and lasers to measure refractive differences. Each cell's refraction is measured and plotted with the entire sample. Plots are used extensively in cell-based research to highlight statistically significant trends within the data.

Flow cytometry was an essential part of two papers published in separate journals by a single lab. The articles, despite reporting unrelated findings, featured the seemingly identical flow cytometry plots seen in Fig. 6.1. The top figure, from the 2010 *Nature* article, purported to show the frequency of blood cells in an experiment involving donor cells from a young mouse being exposed to a specific stem cell niche for 36 h, while the bottom figure, from the 2008 *Blood* article, allegedly represented the frequency of blood cells when wild type donor cells were exposed to a specific stem cell niche for 12 h (Mayack and Wagers 2008; Mayack et al. 2010a). The striking similarity between the two plots was noticed, and immediately after publication in its journal, all of the authors except Mayack retracted the paper, explaining that the reexamination of their work had "undermined [their] confidence in the support for the scientific conclusions reported" (Mayack et al. 2010b). Following the *Nature* retraction, *Blood* voiced serious concerns about Wagers' 2008 publication, which ultimately led to its retraction. In retracting the paper, Wagers stated it was "now confirmed by a subsequent institutional investigation, this paper was found to contain duplicated data and other inappropriate manipulations" (Mayack et al. 2010b). Mayack continued to maintain that the results in both papers were valid and that this was an honest mistake.[43]

Beyond the question of veracity in reporting findings, there are several important ethical implications. First, this case raises a concern about the role of peer review.[44] A more rigorous peer review process may have prevented the publication of the second paper, or required the researchers to produce the "correct" images. Furthermore, the four authors on the 2010 paper should have reviewed the images prior to submission. Image duplication should have been found, as two of the same authors contributed to both works. Second, the *Nature* paper was cited 13 times in the 10 months prior to its retraction; the accuracy of those papers may be in question. Third, the fact that Mayack was an advisee of Wagers highlights some of the challenges in mentorship and authorship; that is, the boundaries of responsibility and credit can easily blur.[45]

Image adjustment software (such programs as Irfanview, Adobe Photoshop, Adobe Illustrator, NIH ImageJ, and commercial software packaged with microscopes, for example, MetaMorph and Nikon Elements) make for endless possibilities of manipulation. The software can be used appropriately to highlight a point the researcher is trying to get across, but could also be used to create false interpretations, remove outliers or anomalies in a biased manner, or make an effect seem more prominent than it may actually be.[46]

For most applications, researchers rely upon monochrome images derived from high-resolution CCD cameras or photomultiplier tubes on their microscopes. For presentation, these images are false-colored using different color maps to represent the fluorophore of interest. For example, in the RGB color map system, a color image is a composite of a red (R), green (G), and blue (B) image. Using a CMYK color map results in an image being a composite of a cyan (C), magenta (M), yellow (Y), and black (K) image. Imaging software allows the user to independently alter the proportion of each color represented in the final composite image as well as the relative intensity of the pixels in each image (e.g., brightness, gamma, contrast). Some such fine-tuning and modification is a matter of aesthetic choice, and can demonstrate the skill of the researcher in presenting a strong visual argument (Frow 2012). But at what point does enhancing an image become inappropriate manipulation?

---

[43] The ORI reviewed the case and found that Dr. Shane Mayack engaged in research misconduct regarding the two papers. The flow cytometry contour plots were found to be falsely represented; additionally, the identical plots contained different labels and numerical percentages (Federal Register, 77(167), 52034–52035; Aug 28, 2012).

[44] Referees often require authors perform further experiments prior to publication, signaling mistrust in the peer review process today. Lack of trust increases costs and causes publication delay (Ploegh 2011).

[45] For example, what if a junior author or a graduate student realizes that coauthors have omitted data that might tell a different story? There are protections for whistleblowing, however, in the reality of day-to-day laboratory environments, relationships with mentors and senior personnel could be compromised, creating considerable risk for the junior researcher.

[46] Thanks to Tom Keenan, Ph.D., for elaborating the following examples.

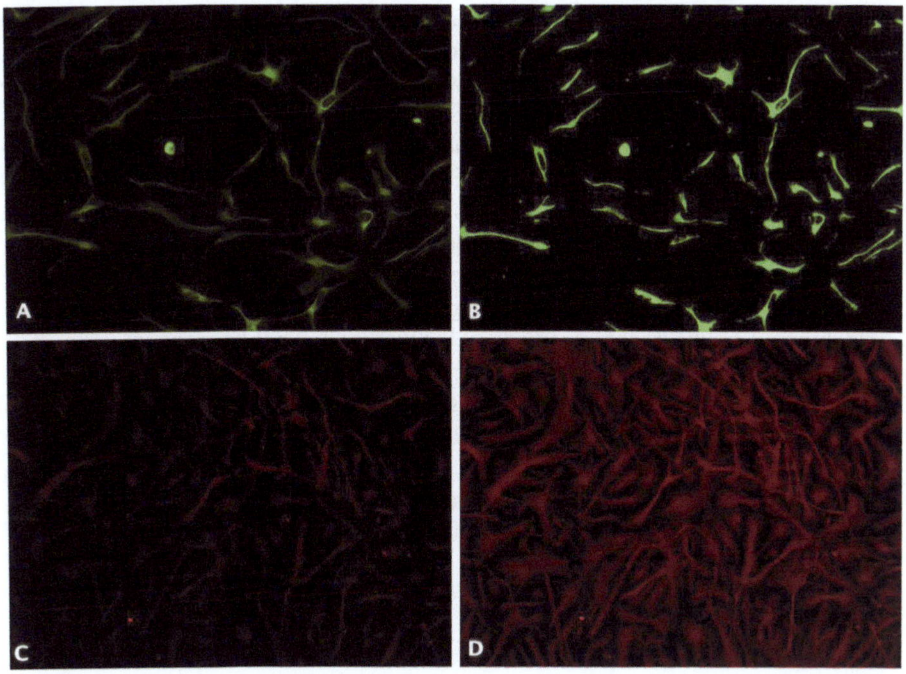

Fig. 6.2 (**a**) The fluorescent microscopy image of cell markers as seen without any image adjustment. (**b**) The same image by disproportionally increasing the *green* component of the image. The figure (**b**) seems to indicate robust expression of the marker. In (**c**), the immunostain image presented seems to show minimal expression of a cell marker. However, this image was taken with a very low gamma setting; correcting the gamma setting shows relatively robust expression as seen in (**d**). Images courtesy of Tom Keenan

If each pixel of the image is adjusted equally there may not be a problem. However, disproportionately adjusting the entire image or component colors of an image can result in the enhancement of particular features over others, or the selective elimination of certain features in the image. Cells can be made to look like they express a molecule when they do not, or not express a molecule when they do. Images can be enhanced to make molecules appear to co-localize when they do not, or to be expressed in specific parts of a cell when they may be expressed more broadly. Image filters can be used to mask undesired parts of an image. One of the most common treatments is to fully saturate the features of an image to make expression of a molecule appear more robust than it truly is. Normally, some structures in a cell would be more or less visible than others, giving the appearance of texture, but when an image is saturated, the appearance of the stain is intensified throughout, and texture disappears. The resulting image appears to show a robust biological response (Fig. 6.2).

Markers that are well documented to demarcate specific cell types or tissue regions may not be as specific as when used in other situations, such as dissociated cell culture.

The distribution of a molecule within certain cells combined with the density of those cells within the tissue can make the marker very specific within a histological section, but more difficult to detect or distinguish when that tissue architecture is lost as in the case of dissociated culture. The researcher may be motivated to accentuate any differences in staining between that cell and other cells within the culture knowing that the marker is well accepted by the greater community (and thus required for publication) even if the expression is not that distinguishable or specific in the original image. The lack of specificity can also occur when researchers use species-specific biomarkers in human samples. For example, researchers may obtain radial glial cell markers for murine cells from a bank; however, using the cell markers in human samples may yield different results, as the specificity of the marker will not be identical.

The bottom line is that the communication of scientific work is based on trust. Peer review should, but does not always, catch problems with data visualization (Nature Editors 2006). Authors must state clearly how figures and images were derived and compiled, and readers need to be able to trust what they see—or be savvy enough to catch red flags, which may be difficult for interdisciplinary or less-experienced researchers.

6.3.4 *Conflicts of Interest*

Research should be conducted without bias from researchers or sponsors who may have an interest in particular outcomes for financial or other reasons. Conflicts may be financial (for example, receiving payments, equity stakes, intellectual property rights, or salary in exchange for expertise or service) or may involve another form of gain, such as influence, access to information, or anything that would create an advantage for a participant. Some activities, such as giving honorariums for writing papers, gifts, or paid travel to give talks, especially when provided by the sponsor of a new product or a politically motivated organization, have long been considered to be routine practice. However, such activities are particularly prone to abuse and have been under increasing scrutiny. In some cases, companies have funded entire departments, multidisciplinary research centers, or campus-wide research programs (Bero 2008). Such corporate sponsorship models often lead to conflicts of interest, as the research institution feels compelled to match its results to its sponsor's needs or withhold data and materials to maintain its sponsor's competitive advantage. As a result of these sources of influence, policy updates were made in 2011 which lowered the threshold for dollar amounts which must be disclosed and changed some reporting requirements to enhance transparency (see Table 6.4).

Additionally, many major journals have started requiring disclosure of both financial and other forms of conflict as a publication requirement. For example, *Nature* requires that its authors disclose any competing financial interests, including funding sources, personal financial interests, and employment, in order to maintain the transparency of the journal and to "help readers form their own judgments of

Table 6.4 Conflict of interest (United States Department of Health and Human Services 2011)[a]

Area	1995 Regulations	2011 Regulations
Significant Financial Interests (SFI)	Investigator must disclose payments and equity interests >$10,000 related to PHS-funded research; income from seminars, lectures, teaching, service on advisory committees or review panels, for public or nonprofit entities exempt from disclosure; travel reimbursement requirements not fully clarified	Investigator must disclose payments and equity interests >$5,000. Reimbursed or sponsored travel related to his/her institutional responsibilities must be disclosed (travel reimbursed or sponsored by federal, state, or local government agency or classes of institutions is exempt from disclosure) Income from seminars, lectures, or teaching, service on advisory or review panels for a federal, state, or local government agency or specific classes of institutions and income from investments investigator does not make direct investment decisions for (mutual funds, retirement accounts, etc.) is exempt
Financial Conflict of Interest (FCOI)	Required to report to PHS institution: grant/contract number, name of principal investigator/project director, name of investigator with FCOI, plan for managing FCOI Institutions must reasonably ensure subrecipient investigators comply	In addition to 1995 Regulations, must describe nature of the FCOI (basis for determination of its existence, how it relates to the research, name of entity with which there is an FCOI), must have management plan including status reports Institutions must ensure their subrecipients comply, including production of a written agreement and timelines Investigator must complete FCOI training before commencing PHS-funded research and at least every 4 years thereafter
Transparency	No regulation	FCOI information (nature of the SFI) available via publicly accessible website or written response upon request
Retrospective review ("Mitigation plan")	No regulation	If noncompliance, institution to conduct retrospective review and notify PHS Awarding Component (full report, including impact on research project, actions institution has or will take to eliminate or mitigate the harm)

[a]This table is adapted from http://grants.nih.gov/grants/policy/coi/summary_of_major_changes.doc. Please consult the original for details and additional information

potential bias."[47] A statement is published at the end of each article disclosing any potential COI. However, COI issues are becoming more prevalent as links between industry and universities change. In 2006, a *Journal of the American Medical Association* survey found that about 67 % of the responding medical school department and large teaching hospital heads had personal relationships with industry (Campbell et al. 2007).

While some activities could clearly be seen as intended to influence scientists, many cases are far more complicated. How far is too far in terms of accepting financial support in a time of squeezed academic budgets and pressure to translate novel research into clinical use? Should researchers reject funding support from for-profit companies because of the potential for influence, even if these may be the only sources to get a promising product into the clinic? It is not uncommon for universities receiving federal funds to enter into agreements with private companies, giving patent rights to the company for university researchers' discoveries in exchange for equity or royalty claims, but it could be argued that this is a conflict for public universities. Another kind of conflict exists if a nondisclosure agreement includes the right of a company to review any publications submitted by the university researchers.

There are also tensions between some federal policies, for example, those which encourage collaboration between academic researchers and industry, and others which appear to discourage such interactions. The Bayh-Dole Act of 1980 (35 U.S.C. §200–212 (1980)) incentivizes inventors to move their products and ideas into the public domain by allowing academic researchers receiving federal funds to retain rights to their intellectual property derived from those public investments. But consider these common scenarios: how should we perceive the actions of an entrepreneurial researcher with an academic appointment who uses university resources for funding, materials, and equipment purchases and then uses them to launch a start-up company? How should we view small start-up companies wishing to collaborate with nearby universities in alignment with local or state economic development incentives, and then uses academic personnel (including student trainees) to do things they cannot afford to do? Are these instances of potential COI governable by RCR rules, or reflections of contemporary science as driven by legislative actions (such as Bayh-Dole) and market-driven translational science as promulgated by the NIH, the MRC in the UK, and other international funding bodies? When researchers are evaluated (and promoted) based as much on the number of patents and commercial ventures they engage in as the quality of their teaching and research, what is the relationship of COI rules with other institutional priorities?

In addition to significant financial conflicts, there may also be conflicts of commitment (percent effort on a project or other obligations), intellectual bias (for example, any relationship or activity which might negatively affect peer review of

[47] Loewenstein argues that disclosures of COI may be harmful, for example, if the discloser attempts to overstate a position in order to overcome a predicted discount of his or her advice, or when disclosers feel they can provide biased information because they disclosed the conflict. Another type of harm exists when patients decide not to enroll in certain beneficial studies because of their doctor's disclosed affiliations (Loewenstein et al. 2012).

grants or publications), or conscience (when one's personal beliefs affect objectivity, for example, if a reviewer rejected work involving derivation of embryonic stem cells or a university leader suppressed research on embryonic stem cells due to personal beliefs about embryos and embryo use).

6.4 Discussion

In this chapter, we have provided an overview of existing RCR components, where they came from, how they are administered, and potential consequences of noncompliance. We also demonstrated the kinds of situations which may fall into gray areas, or which may not yet have been adequately addressed by current checkboxes. There are many helpful sources of information regarding scientific integrity, but researchers must not only be aware of research ethics issues, they must learn to discern when something goes beyond ethically acceptable practice, and be proactive in seeking assistance to resolve potential problems.

We have also demonstrated how both the dynamic nature of scientific inquiry and social, political, and economic contingencies may shape RCR oversight systems and affect the interpretation and implementation of rules (or not). Existing essential components of RCR training capture core concerns of scientific integrity; however, as we have shown, a number of things pose questions not yet adequately addressed inside RCR boxes. These include technical issues such as the emergence of new research tools and techniques, and the capability of creating novel life forms, among other things. For example, tools such as imaging software, FACS, gene sequencing, and others have dramatically expanded research capabilities, yet as we have shown, many of these techniques can be exploited to enhance or misrepresent findings. The ability to create chimeras moves research ethics debates beyond the suffering of laboratory animals and disturbs assumed distinctions between species, causing us to rethink what counts as "human." Likewise, the "human" in human biological materials makes it difficult to dissociate biological from social identities, even when labels are removed, creating confusion about exactly what or who we are protecting as a "human subject."

But there are also broader social issues affecting the conduct of science, such as changing expectations of academic researchers and students in an era of pressure to produce translational research, shifting legal definitions and social understandings of privacy and protections, and novel social arrangements (including academic–industry–state relations and methods of supporting research), among other things. For example, the proposed change in human subjects protections is in part related to new capabilities arising from the "omics" sciences, as we have discussed, but also speaks to the unsettled meanings and yet importance of privacy in society, as well as broader issues such as the tensions between the common good and the protection of individuals' interests.

Social conditions also affect views of how much oversight is needed, and where responsibility lies. During the era of "downsizing government" initiatives in the

1990s, oversight functions were diminished. More recently, there has been an increase in mandates at the federal level, including requirements for specific RCR training. Unfortunately, there has been no corresponding investment to support the execution of said mandates, forcing universities and research institutions to take resources from other areas. Inevitably, critics of enhanced oversight will argue that the additional bureaucracy constrains science and increases costs while decreasing productivity, while proponents will point to the high economic and human costs of not having sufficient protections in place. A different route would be neither more nor less oversight, but smarter oversight; that is, considering carefully what is the real problem underlying the oversight requirement being proposed, and if there are alternative ways to prevent problems. A beginning step would be to recognize that misconduct is not simply a matter of individual "bad apples;" rather, there are systemic issues that should be addressed, such as regulations and guidelines that contradict or duplicate each other, agencies which have differing priorities regarding RCR, or legal mechanisms working at cross purposes, for example, those that attempt to protect intellectual property or control resource use while simultaneously promising donors of tissue or inventors of techniques control over disposition. Conflicting social, political, and economic demands also shape the RCR landscape. For example, the present push toward translating research into commercial products comes with an emphasis on removing barriers (such as informed consent or privacy protections, which may be seen to interfere with the use of human subjects or data from them), which is in tension with the priority American bioethics places on autonomy and privacy, and a long-standing repulsion at the notion of commercializing human bodily materials.

Debates will always exist about whether certain novel technologies require special, additional oversight. Currently, there is no systematic guidance on when a novel technology calls for additional oversight (especially to protect human subjects) and when ordinary oversight suffices. A recent example is nanotechnology. The most beneficial aspect of nanomedicine is that materials can be made to change properties in situ in the human body. This becomes the greatest disadvantage, because emergent properties may have unpredictable effects. Scientists were also alarmed about unknowable effects of nanoparticle exposure to workers and the environment. One analysis recommended a middle road which included closer coordination among federal agencies through an interagency working group, and a Secretary level committee to discern when particular nanomedicine innovations required additional scrutiny for oversight (Fatehi et al. 2012). Similarly, concern over the use of synthetic biology techniques accelerated once a synthetic genome was successfully integrated into a bacterial cell, creating an organism not found in nature. A report from the Presidential Commission for the Study of Bioethical Issues recommended "prudent vigilance" with federal oversight of the research (PCSBI 2010).[48] The 18 recommendations recognized that due to the capability of conducting synthetic biology research on small scales and in nontraditional settings, much of this research may proceed in smaller labs which may not be accustomed to

[48] http://bioethics.gov/cms/synthetic-biology-report.

complying with RCR requirements. The PCSBI called for required education on the ethical dilemmas raised by synthetic biology, but did not advocate for an overhaul of existing regulations. The PCSBI report was contested by a group of more than 100 nongovernmental organizations, which felt that greater precaution should be exercised.

Short of reinventing or adding to oversight systems, other recommendations might include greater coordination among federal agencies (especially given the differing histories and legislative authorities under which various RCR requirements exist), mechanisms such as independent fact-checking resources to verify claims made in published papers or reports, or better tools with which to detect misrepresented findings. An innovation some universities are adopting is a "bench-side consulting service," through which anyone can bring questions raised in the course of research to a group of local experts, usually a collection of scientists, ethicists, or policy experts, who can advise them safely and discreetly before commencing the research. We invite researchers, ethicists, and policy makers to "think beyond the checkboxes" of research ethics to address some of the challenges we raise here, and to consider how best to responsibly conduct research.

References

Aalto-Setälä, K., Conklin, B. R., & Lo, B. (2009). Obtaining consent for future research with induced pluripotent cells: Opportunities and challenges. *PLOS Biology, 7*(2), e42.

Academy of Medical Sciences. (2011). *Animals containing human material*. London: The Academy of Medical Sciences.

American Association of Blood Banks. (2009). Circular of information for the use of cellular therapy products. Retrieved October 5, 2012, from http://www.aabb.org/resources/bct/Documents/coi_ct1109.pdf.

American Society for Reproductive Medicine. (2009). Donating spare embryos for stem cell research. *Fertility and Sterility, 91*(3), 667–670. doi:10.1016/j.fertnstert.2008.12.066.

Annas, G. J., & Grodin, M. A. (1995). *The Nazi doctors and the Nuremberg Code: Human rights in human experimentation*. Oxford: Oxford University Press.

Baggerly, K. A., & Coombes, K. R. (2010). Deriving chemosensitivity from cell lines: Forensic bioinformatics and reproducible research in high-throughput biology. *Annals of Applied Statistics, 3*(4), 1309–1334. doi:10.1214/09-AOAS291.

Bahadur, G., Iqbal, M., Malik, S., Sanyal, A., Wafa, R., & Noble, R. (2008). Admixed human embryos and stem cells: Legislative, ethical and scientific advances. *Reproductive Biomedicine Online, 17*(Suppl 1), 25–32.

Baylis, F. (2008). Animal eggs for stem cell research: A path not worth taking. *The American Journal of Bioethics, 8*(12), 18–32. doi:10.1080/15265160802559161.

Baylis, F., & McLeod, C. (2007). The stem cell debate continues: The buying and selling of eggs for research. *Journal of Medical Ethics, 33*(12), 726–731. doi:10.1136/jme.2007.022129.

Bero, L. (2008). "Experimental" institutional models for corporate funding of academic research: Unknown effects on the research enterprise. *Journal of Clinical Epidemiology, 61*(7), 629–633.

Bobrow, M. (2011). Regulate research at the animal-human interface. *Nature, 475*(7357), 448.

Campbell, E. G., Weissman, J. S., Ehringhaus, S., Rao, S. R., Moy, B., Feilbelmann, S., et al. (2007). Institutional academic-industry relationships. *Journal of the American Medical Association, 298*(15), 1779–1786. doi:10.1001/jama.298.15.1779.

Caulfield, T., Zarzeczny, A., McCormick, J., Bubela, T., Chritchley, C., Einsiedel, E., et al. (2009). International stem cell environments: A world of difference. *Nature Reports Stem Cells*. doi:10.1038/stemcells.2009.61.

Corbyn, Z. (2012). Misconduct is the main cause of life-sciences retractions. *Nature, 490*(7418), 21. doi:10.1038/490021a.

Couzin-Frankel, J. (2010). Cancer research: As questions grow, Duke halts trials, launches investigation. *Science, 329*(5992), 614–615.

Cyranoski, D. (2012). Stem-cell therapy takes off in Texas. *Nature, 483*(7387), 13–14.

Daley, G. Q., Ahrlund Richter, L., Auerbach, J. M., Benvenisty, N., Charo, R. A., Chen, G., et al. (2007). Ethics. The ISSCR guidelines for human embryonic stem cell research. *Science, 315*(5812), 603–604. doi:10.1126/science.1139337.

Department of Health and Human Services. (2011). Human subjects research protections: Enhancing protections for research subjects and reducing burden, delay, and ambiguity for investigators. Resource document. Health and Human Services, Food and Drug Administration. Retrieved, September 1, 2012, from http://www.gpo.gov/fdsys/pkg/FR-2011-07-26/pdf/2011-18792.pdf.

Deutscher Ethikrat (2011). *Mensch-Tier-Mischwesen in der Forschung Stellungnahme*. Berlin: Deuscher Ethikrat.

Downs, K. (2008). Embryological origins of the human individual. *DNA and Cell Biology, 27*(1), 3–7. doi:10.1089=dna.2007.0698.

Elger, B., & Caplan, A. (2006). Consent and anonymization in research involving biobanks. *EMBO Reports, 7*(7), 661–666.

Elster, A., Damaschun, A., Kurtz, A., Stacey, G., Arán, B., Veiga, A., et al. (2008). The changing landscape of European and international regulation on embryonic stem cell research. *Stem Cell Research, 2*, 101–107. doi:10.1016/j.scr.2008.10.003/.

European Research Council. (2012). ERC Scientific Misconduct Strategy. Retrieved, March 12, 2013, from http://erc.europa.eu/sites/default/files/document/file/ERC_Scientific_misconduct_strategy.pdf.

Fanelli, D. (2010). Do pressures to publish increase scientists' bias? An empirical support from US States' Data. *PLOS ONE., 5*, e10271. doi:10.1371/journal.pone.0010271.

Fang, F. C., Steen, R. G., & Casadevall, A. (2012). Misconduct accounts for the majority of retracted scientific publications. *PNAS, 109*(42), 17028–17033. doi:10.1073/pnas.1212247109.

Fatehi, L., Wolf, S. M., McCullough, J., Hall, R., Lawrenz, F., Kahn, J. P., et al. (2012). Recommendations for nanomedicine human subjects research oversight: An evolutionary approach for an emerging field. *Journal of Law, Medicine and Ethics, 40*(4), 716–750.

Fisher, J. A. (2005). *Pharmaceutical paternalism and the privatization of clinical trials*. Troy: Rensselaer Polytechnic Institute.

Frow, E. K. (2012). Drawing a line: Setting guidelines for digital image processing in scientific journal articles. *Social Studies of Science*. doi:10.1177/0306312712444303.

Gottweis H. (2002). Stem Cell Policies in the United States and in Germany. *Policy Studies Journal*. 30(4), 444–469.

Gottweis, H., & Triendl, R. (2006). South Korean policy failure and the Hwang debacle. *Nature Biotechnology*. doi:10.1038/nbt0206-141.

Greely, H. T. (2013). Assessing ESCRs: Yesterday and tomorrow. *American Journal of Bioethics, 13*(1), 44–52. doi:10.1080/15265161.2013.747340.

Greely, H. T., Cho, M. K., Hogle, L. F., & Satz, D. M. (2007). Thinking about the human neuron mouse. *The American Journal of Bioethics, 7*(5), 27–40.

Gunter, K. C., Caplan, A. L., Mason, C., Salzman, R., Janssen, W. E., et al. (2010). Cell therapy medical tourism: Time for action. ISCT White Paper. *Cytotherapy, 12*, 965–968.

Gymrek, M., McGuire, A. L., Golan, D., Halperin, E., & Erlich, Y. (2013). Identifying personal genomes by surname inference. *Science, 339*, 321–324.

Haber, M. H., & Benham, B. (2012). Reframing the ethical issues in part-human animal research: The unbearable ontology of inexorable moral confusion. *The American Journal of Bioethics, 12*(9), 17–25.

Hansson, M. G., Dillner, J., Bartram, C. R., Carlson, J. A., & Helgesson, G. (2006). Should donors be allowed to give broad consent to future biobank research? *The Lancet Oncology, 7*(3), 266–269.

Hogle, L. F. (2011). *Anonymity and identity at the cellular level.* Paper presented to the American Anthropological Association, Toronto, Canada.

Hyun, I., Lindvall, O., Ahrlund-Richter, L., Cattaneo, E., Cavazzana-Calvo, M., Cossu, G., et al. (2008). New ISSCR guidelines underscore major principles for responsible translational stem cell research. *Cell Stem Cell, 3*(6), 607–609. doi:10.1016/j.stem.2008.11.009.

Institute of Medicine. (2012). *Evolution of translational omics: Lessons learned and the path forward.* Washington, DC: National Academies Press.

International Society for Stem Cell Research. (2006). Guidelines for the conduct of human embryonic stem cell research. Retrieved, October 27, 2012, from http://www.isscr.org/docs/default-source/hesc-guidelines/isscrhescguidelines2006.pdf.

International Society for Stem Cell Research. (2008).Guidelines for the clinical translation of stem cells. Retrieved, October 27, 2012, from http://www.isscr.org/docs/default-source/clin-trans-guidelines/isscrglclinicaltrans.pdf.

Isasi, R. & Knoppers, B. (2006). Mind the gap: policy approaches to embryonic stem cell and cloning research in 50 countries. *European Journal of Health Law*, 13(1), 9–25.

Isasi, R., Knoppers, B., & Lomax, G. (2011). Sustained interaction: The new normal for stem cell repositories? *Regenerative Medicine, 6*(6), 783–792. doi:10.2217/rme.11.93.

Ishikawa, K., Takenaga, K., Akimoto, M., Koshikawa, N., Yamaguchi, A., Nakada, K., et al. (2008). ROS-generating mitochondrial DNA mutations can regulate tumor cell metastasis. *Science, 320*(5876), 661–664.

Jonsen, A. (1998). *The birth of bioethics*. New York: Oxford University Press.

Karpowicz, P., Cohen, C., & van der Kooy, D. (2005). Developing human-nonhuman chimeras in human stem cell research: Ethical issues and boundaries. *Kennedy Institute of Ethics Journal, 15*(2), 107–134.

Katnelson, A. (2010). Stem-cell papers under suspicion. *Nature News*. doi:10.1038/news.2010.544.

Kawakami, M., Sipp, D., & Kato, K. (2010). Regulatory impacts on stem cell research in Japan. *Cell Stem Cell, 6*(5), 415–418.

Knowles, L. (2003). Ethics of using hybrids, chimeras and cytoplasmic hybrids. *Stem Cell Network (Canada).* Retrieved, from http://www.stemcellnetwork.ca/uploads/File/whitepapers/Ethics-of%20Research-Using-Hybrids.pdf

Krueger, J. (2009). Incidences of ORI cases involving falsified images. *Office of Research Integrity Newsletter, 17*, 2–3.

Laflamme, M. A., Chen, K. Y., Naumova, A. V., et al. (2007). Cardiomyocytes derived from human embryonic stem cells in pro-survival factors enhance function of infarcted rat hearts. *Nature Biotechnology*. doi:10.1038/nbt1327.

LaFollette, M. C. (2000). The evolution of the "scientific misconduct" issue: An historical overview. *Proceedings of the Society for Experimental Biology and Medicine, 224*, 211–215.

Lederer, S. (1997). *Subjected to science: Human experimentation in America before the Second World War.* Baltimore: Johns Hopkins University Press.

Lederer, S. E. (2008). *Flesh and blood: Organ transplantation and blood transfusion in 20th century America.* New York: Oxford University Press.

Lee, J. B., & Park, C. (2006). Molecular genetics: Verification that Snuppy is a clone. *Nature, 440*, E2–E23.

Lensch, M. W., Schlaeger, T. M., Zon, L. I., & Daley, G. Q. (2007). Teratoma formation assays with human embryonic stem cells: A rationale for one type of human-animal chimera. *Cell Stem Cell, 1*(3), 253–258. doi:10.1016/j.stem.2007.07.019.

Levine, R. J. (1988). *Ethics and regulation of clinical research.* New Haven: Yale University Press.

Levine, A. D. (2011). Policy uncertainty and the conduct of stem cell research. *Cell Stem Cell, 8*(2), 132–135. doi:10.1016/j.stem.2011.01.002.

Lo, B., Parham, L., Broom, C., Cedars, M., Gates, E., Guidice, L., et al. (2009). Importing human pluripotent stem cell lines derived at another institution: Tailoring review to ethical concerns. *Cell Stem Cell, 4*(2), 115–123.

Loewenstein, G., Sah, S., & Cain, D. M. (2012). The unintended consequences of conflict of interest disclosure. *Journal of the American Medical Association, 307*(7), 669–670.

Lomax, G., & Stayn, S. (2008). Similarities and differences among stem cell research policies: Opportunities for policymakers, patients, and researchers. *Medical Research Law & Policy Report, 7*(21), 695–698.

Lowenthal, J., Lipnick, S., Rao, M., & Hull, S. (2012). Specimen collection for induced pluripotent stem cell research: Harmonizing the approach to informed consent. *Stem Cells Translational Medicine, 1*, 409–421.

Maher, B. (2012). The new gatekeepers: Reducing research misconduct. *Nature News Blog*. Retrieved, September 1, 2012, from http://blogs.nature.com/news/2012/03/the-new-gatekeepers-reducing-research-misconduct.html.

Marks, H. (1997). *The progress of experiment: Science and therapeutic reform in the United States, 1900–1990*. Cambridge: Cambridge University Press.

Marris, E., & Check, E. (2006). Disgraced cloner's ally is cleared of misconduct. *Nature, 439*(7078), 768–769. doi:10.1038/439768b.

Martinson, B. C., Anderson, M. S., & de Vries, R. (2005). Scientists behaving badly. *Nature, 435*, 737–738.

Mayack, S. R., Shadrach, J. L., Kim, F. S., & Wagers, A. J. (2010a). Systemic signals regulate ageing and rejuvenation of blood stem cell niches. *Nature, 463*, 495–500. doi:10.1038/nature08749.

Mayack, S. R., Shadrach, J. L., Kim, F. S., & Wagers, A. J. (2010b). Systemic signals regulate ageing and rejuvenation of blood stem cell niches. *Nature, 467*, 872. doi:10.1038/nature09474 (Retraction).

Mayack, S. R., & Wagers, A. J. (2008). Osteolineage niche cells initiate hematopoietic stem cell mobilization. *Blood, 112*(3), 519–531. doi:10.1182/blood-2008-01-133710.

McAllister, T. N., Audley, D., & L'Heureux, N. (2012). Autologous cell therapies: Challenges in US FDA regulation. *Regenerative Medicine, 7*(6 Suppl), 94–97.

McGuire, A. L., Caulfield, T., & Cho, M. K. (2008). Research ethics and the challenge of whole-genome sequencing. *Nature Reviews Genetics, 9*, 152–156. doi:10.1038/nrg2302.

McMahon, D. S., Thorsteinsdottir, H., Singer, P. A., & Daar, A. S. (2010). Cultivating regenerative medicine innovation in China. *Regenerative Medicine, 5*(1), 35–44.

Moreno, J. (2001). *Undue risk: Secret state experiments on humans*. New York: Routledge.

Muotri, A. R., Nakashima, K., Toni, N., Sandler, V. M., & Gage, F. H. (2005). Development of functional human embryonic stem cell-derived neurons in mouse brain. *Proceedings of the National Academy of Sciences of the United States of America, 102*(51), 18644–18648. doi:10.1073/pnas.0509315102.

National Academies of Science. (1996). *Guide for the care and use of laboratory animals*. The National Academies Press. Retrieved, October 24, 2012, from http://www.nap.edu/openbook.php?record_id=5140.

National Academies of Science. (2005). *Guidelines for human embryonic stem cell research*. Washington, DC: The National Academies Press.

National Institutes of Health. (2009). Update on the requirement for instruction in the responsible conduct of research. Notice NOT-OD-10-019. Retrieved, September 1, 2012, from http://grants.nih.gov/grants/guide/notice-files/not-od-10-019.html.

Nature Editors. (2006). Peer review and fraud. *Nature*, 444, 971–972. doi: 10.1038/444971b.

Normile, D. (2009). Japan streamlines embryonic stem cell reviews. American association for the advancement of science. Retrieved, January 20, 2013, from http://news.sciencemag.org/scienceinsider/2009/08/japan-streamlin.html.

Normile, D., Vogel, G., & Couzin, J. (2006). South Korean team's remaining human cell claim demolished. *Science*. doi:10.1126/science.311.5758.156.

Office for Human Research Protections. (2013). International compilation of human research standards. Retrieved, from http://www.hhs.gov/ohrp/international/intlcompilation/intlcomp2013.pdf.pdf.

Office of Science and Technology Policy. (2000). Federal policy on research misconduct. *Federal Register, 65*(235), 76260–76264.

Petryna, A. (2007). Clinical trials offshored: On private sector science and public health. *BioSocieties, 2*, 21–40.
Ploegh, H. (2011). End the wasteful tyranny of reviewer experiments. *Nature, 472*(7344), 391. doi:10.1038/472391a.
Presidential Commission for the Study of Bioethical Issues. (2010). *Report*: *The ethics of synthetic biology and emerging technologies*. Washington, DC. Retrieved, from http://bioethics.gov/cms/sites/default/files/PCSBI-Synthetic-Biology-Report-12.16.10_0.pdf
Reverby, S. M. (2009). *Examining Tuskegee: The Infamous Syphilis Study and its legacy*. Chapel Hill: University of North Carolina Press.
Robert, J. S., & Baylis, F. (2003). Crossing species boundaries. *The American Journal of Bioethics, 3*(3), 1–13. doi:10.1162/15265160360706417.
Russell, W. M. S., & Burch, R. L. (1959). *The principles of humane experimental technique*. London: Metheun & Co. Ltd.
Safadi, R., & Pappo, O. (2007). Amelioration of hepatic fibrosis via beta-glucosylceramide-mediated immune modulation is associated with altered CD8 and NKT lymphocyte distribution. *International Immunology, 19*(8), 1021–1029.
Safadi, R., & Pappo, O. (2011). Retraction of: Amelioration of hepatic fibrosis via beta-glucosylceramide-mediated immune modulation is associated with altered CD8 and NKT lymphocyte distribution. *International Immunology*. doi:10.1093/intimm/dxr0543.
Saunders, R., & Savulescu, J. (2008). Research ethics and lessons from Hwang-gate: What can we learn from the Korean cloning fraud? *Journal of Medical Ethics, 34*, 214–221. doi:10.1136/jme.2007.023721.
Scott, R., Williams, C., Ehrich, K., & Farsides, B. (2012). Donation of 'spare' fresh or frozen embryos to research: Who decides that an embryo is 'spare' and how can we enhance the quality and protect the validity of the consent? *Medical Law Review, 20*(3), 255–303.
Siegel, A. W. (2003). The moral insignificance of crossing species boundaries. *The American Journal of Bioethics, 3*(3), 33–34. doi:10.1162/15265160360706525.
Snyder, E. Y., & Loring, J. F. (2006). Beyond fraud: Stem-cell research continues. *New England Journal of Medicine, 354*, 321–324.
Streiffer, R. (2005). At the edge of humanity: Human stem cells, chimeras, and moral status. *Kennedy Institute of Ethics Journal, 15*(4), 347–370.
Streiffer, R. (2008). Informed consent and federal funding for stem cell research. *Hastings Center Report, 38*(3), 40–47.
Taylor, P. L. (2009). Retroactive ethics in rapidly developing scientific fields. *Cell Stem Cell, 4*(6), 479–482. doi:10.1016/j.stem.2009.05.002.
The Hinxton Group. (2006). Transnational cooperation in stem cell research. Consensus statement. Retrieved, from http://www.hinxtongroup.org/docs/Hinxton%202006%20consensus%20document.pdf.
United States Department of Health and Human Services. (2005). Public health services research polices on scientific misconduct. *Federal Register, 70*(94), 28370–28400.
Van Noorden, R. (2011). Science publishing: The trouble with retractions. *Nature, 478*(7367), 26–28. doi:10.1038/478026a.
Werner, M., Mayleben, T., & Van Bokkelen, G. (2012). Autologous cell therapies: The importance of regulatory oversight. *Future Medicine, 7*(6s), 100–103.
Winickoff, D. E., Saha, K., & Graff, G. D. (2009). Opening stem cell research and development: A policy proposal for the management of data, intellectual property, and ethics. *Yale Journal of Health Policy Law & Ethics, 52*, 54–127.
World Medical Association. (1975). Declaration of Helsinki - Ethical Principles for Medical Research Involving Human Subjects. Retrieved, April 12, 2013, from http://www.wma.net/en/30publications/10policies/b3/17c.pdf.
Zhang, J. Y. (2012). The de-nationalization and re-nationalization of the life sciences in China: A cosmopolitan practicality? *Foundation Maison des Sciences de l'Homme—White Paper*—FMSH-WP-2012-08.

Chapter 7
Protecting Human Participants in the Procurement of Materials in Regenerative Medicine Research

David B. Resnik

7.1 Introduction

Human participants provide a variety of materials that are essential for research in regenerative medicine, including adult stem cells, embryos, sperm, ova, blood, and urine. Investigators use these materials in their own research, share them with other investigators, or deposit them in biobanks for future use. Because regenerative medicine depends so heavily on the cooperation of human participants, researchers must take appropriate measures to protect their rights and welfare and secure their trust. If participants cannot trust researchers to protect them from harm or exploitation, they will not donate materials. This chapter will focus on some emerging issues in the procurement of materials in regenerative medicine research that merit special attention.

Many of the ethical issues that arise in regenerative medicine research with human participants are similar to those that occur in other types of biomedical research. However, regenerative medicine researchers face some problems not typically encountered by other biomedical researchers, due to the nature of their subject matter. To understand some of these unique problems, it is important to reflect on (1) the types of materials provided by participants, (2) the uses of these materials, and (3) sources of materials.

Concerning the types of materials procured, a key distinction is between (A) adult (or somatic) stem cells and other nonreproductive materials and (B) embryos and other reproductive materials, such as sperm or ova. Researchers must keep this distinction in mind when they procure biological samples from participants, because these types of materials have different implications for the rights and welfare of donors. The ethical issues of acquiring adult stem cells (e.g., cells taken from skin,

D.B. Resnik, J.D., Ph.D. (✉)
National Institute of Environmental Health Sciences, National Institutes of Health,
Research Triangle Park, CA, USA
e-mail: resnikd@niehs.nih.gov

L.F. Hogle (ed.), *Regenerative Medicine Ethics: Governing Research and Knowledge Practices*, DOI 10.1007/978-1-4614-9062-3_7,

fat, muscle, or bone marrow biopsies) are similar, in principle, to those of any other type of nonreproductive human tissue, because adult stem cells do not normally have the potential to become human beings. Adult stem cells can be reprogrammed to become induced pluripotent stem cells (cells that can differentiate into any tissue type). Although it is theoretically possible to transform human adult stem cells into precursors of sperm or ova, this is not a well-tested or common procedure.

Embryos, however, can become human beings if implanted in the womb and gestated successfully, and sperm and ova can unite in vitro to form embryos. These procedures are commonly used in fertility clinics to help couples produce children. Thus, issues of human reproduction should always be considered when acquiring reproductive materials, though they need not normally be addressed when acquiring adult stem cells or other nonreproductive materials. Acquiring embryos leftover from in vitro fertilization attempts therefore raises issues concerning their use, storage, transfer, and destruction that investigators must address. Individuals may think differently about the use of their embryos in research than they do about the use of their skin or fat cells, because they may regard embryos as having a special moral status related to their potential to become human beings. Destruction of embryos, which occurs in some types of procedures that harvest embryonic stem cells, raises concerns for those who regard embryos as having a special moral status. Couples may also want to be assured that their embryos will not be cloned or used for reproductive purposes by someone else (Cohen 2007).

The procurement of oocytes and sperm raises concerns similar to the procurement of embryos, because researchers can fertilize ova in vitro to form embryos. Additionally, eggs can be used in a type of cloning known as somatic cell nuclear transfer (SCNT), in which the nucleus from an egg is removed and replaced with the nucleus from a donor cell. Researchers can use the product of SCNT as a source of stem cells or to produce a human being. Although human reproductive cloning is illegal in many countries, many allow cloning for research or therapeutic purposes (see Appendix A). Although the NIH does not fund research involving the creation of embryos for scientific or medical purposes, some private companies and foundations fund this type of research. The NIH will only fund research on embryos leftover from in vitro fertilization attempts (National Institutes of Health 2009). While this chapter will focus on the procurement of human embryos leftover from in vitro fertilization attempts, researchers should also be mindful of issues related to the procurement of human gametes if they work with these materials (Cohen 2007).

The uses of human biological materials in regenerative medicine research are varied. Some types of uses, such as culturing embryonic stem cells or adult stem cells to study basic cellular mechanisms, raise no unique ethical concerns. However, other uses do. Researchers may implant human cells or tissues into animals to form chimeras, which can be used to better understand physiological mechanisms and to develop animal models for human diseases. The production of chimeras raises ethical concerns because some people may not want to donate materials for this purpose since they may regard human-animal combinations as morally objectionable. Stem cells, or organs or tissues derived from stem cells, may also be implanted in patients to prevent or treat diseases. This use of stem cells would be similar to organ/tissue transplantation. It seems reasonable that donors should be told if their cells will be used for these purposes because some people may not want their tissues growing inside another person. Stem cells may also be used to test toxic chemicals or for commercial purposes, and some people may not want their donated materials used in this way. Stem cells may also be used for large-scale genome sequencing, which raises issues of privacy and confidentiality that most people would be concerned about (Aalto-Setälä et al. 2009).

Regenerative medicine materials may be procured from a variety of sources. Adult stem cells may be procured from adults, children, or aborted fetuses or from the umbilical cord or placenta. Couples may also donate embryos for use in regenerative medicine research. The source of the materials used in research has implications for the consent process.

7.2 Informed Consent Issues

7.2.1 Background

Obtaining informed consent from donors is a key issue in regenerative medicine research.

Regulations and ethical guidelines require that investigators obtain the informed consent of the human participant or the participant's legal representative (Department of Health and Human Services 2009; World Medical Association 2008; Capron 2008). Informed consent is important in research for several reasons. First, allowing individuals to make an informed choice to participate in research respects their autonomy and personal dignity. Some of the worst abuses of human research participants have involved coercion, manipulation, or other violations of autonomy (Shamoo and Resnik 2009). Second, informed consent can help protect individuals from harm by giving them an opportunity to decide whether they want to take risks associated with the research. Although investigators and oversight committees (such as institutional review boards [IRBs]) are required to take steps to protect individuals from harm, participants can also play an important role in safeguarding their own safety and welfare. Third, informed consent can help build trust among investigators and participants, especially if consent involves an extended discussion of the research and is more than just signing a form (Brock 2008).

Before the 1990s, investigators routinely acquired human tissues and other biological samples leftover from clinical procedures without consent (Skloot 2010). One of the most widely used cell lines, known as HeLa, was derived from cervical cancer cells taken from Henrietta Lacks while she was a patient at Johns Hopkins Hospital. HeLa has played a vital role in basic and applied biomedical research. Over 60,000 articles have been published reporting result of research with HeLa. Ms. Lacks did not consent to the use of her tissue and was not informed that researchers had developed a cell line from her tissue. Furthermore, although researchers gained reputation and considerable income from the innovation made from her cells, she did not share in any of the profits (Skloot 2010).

In 1976, researchers developed a cell line from cancer tissue removed from John Moore's body when he had a splenectomy to treat hairy cell leukemia at the University of California at Los Angeles Medical Center. The investigators patented the cell line and licensed it to a private company for hundreds of millions of dollars. When Moore discovered that the researchers had developed and patented the cell line without his permission, he sued the researchers and the university, claiming that he had property rights over his body and materials removed from it. The question of disposition of bodily materials, and in particular, whether or not the body or parts of the body can have a property-like status, has long been in dispute, but this case forced the courts to confront the issues directly. In 1990, the California Supreme Court ruled that, while Moore did not have property rights pertaining to the cell line, the investigators had violated their fiduciary responsibility to Moore as a patient, by not informing him of the intended use of his tissue or obtaining his consent to use it in the way they intended (Resnik 2004). Although the Moore case did not establish property rights for tissue donors, it set a legal precedent for a duty to obtain informed consent for tissue donation, which has been recognized by other courts (Resnik 2004). Today, it is widely acknowledged that informed consent is essential to tissue procurement (Weir and Olick 2004).

7.2.2 Approaches to Consent

During the consent process, investigators should inform participants about the nature of the research, including the procedures, risks, benefits, alternatives, and confidentiality protections. They should also inform participants how to stop participating in a study and withdraw samples or data. It may not be possible to withdraw data that have already been used in research, but samples still can be withdrawn, as long as researchers retain information that links individuals to their donated samples. Investigators should also ask participants if they would like to be contacted about future studies or receive individualized tests results related to their samples or data.[1] This is particularly important with cells which will be immortalized in lines, as they may be used for long periods of time.

An analysis of informed consent forms for human embryonic stem cell (HESC) lines approved by the Bush Administration for use in federally funded research found that many of the documents did not provide donors with adequate information or included clauses that would prohibit some types of legitimate research. For example, some documents said that the embryos would be used only in certain research on specific diseases and some documents did not inform donors that their embryos

[1] The return of individualized tests results, such as genetic or genomic information or clinical laboratory findings, is a controversial topic in bioethics that will not be explored in depth in this chapter. For further discussion, see Wolf et al. (2008), and Resnik (2011).

might be destroyed to harvest embryonic stem cells. As a result, many of the documents drafted in the early 1990s were considered to be inadequate under current ethical standards (Streiffer 2008). Thus, it is important for investigators to pay close attention to informed consent issues pertaining sharing and using biological samples, because failure to adequately inform participants about how their materials may be used or shared may violate autonomy or preclude important types of research.

One of the most important consent issues is providing participants with information about how their samples will be used (Weir and Olick 2004). There are three basic approaches to this topic. According to the specific consent approach, participants should give explicit permission for each use of their samples. Although this approach maximizes participant autonomy, it can significantly impede the sharing and use of samples, because the investigator must contact participants and obtain their consent each time samples will be shared or used in a manner not describe in the original consent form. Also, the specific consent approach may inconvenience participants, who do not want to be bothered by continual requests to use or share samples (Petrini 2010).

As an alternative to the specific consent approach, many investigators use the general (or blanket) consent approach, in which participants give permission for general use of their samples. Granting permission for general use may be an appropriate option for the procurement of nonreproductive materials, such as adult stem cells (International Society for Stem Cell Research 2006; National Academy of Sciences 2005; Isasi et al. 2011). While the general consent approach facilitates the sharing and use samples, it may compromise informed consent. Critics have argued that since participants may not understand the implications of giving general permission for the use of their samples, the validity of their consent is questionable. However, proponents of the general consent approach respond that most research participants understand what it means to give general permission for the use of their samples/data and that they are comfortable with this decision (Wendler 2006).

Although the general consent approach may be appropriate for some types of biomedical research, it probably should not be used in procuring embryos or adult stem cells used in regenerative medicine research, due to the nature of these materials and their special uses. To provide full autonomy, couples who are donating embryos for research should be informed about how their embryos will be used (e.g., to derive embryonic stem cells), shared, and stored. They should be informed if any research procedures will involve the destruction of embryos. Couples should also be informed how long their embryos may be kept in storage. They should be able to refuse to let investigators use their embryos in research if they change their minds. Although research on cell lines derived from the embryos could continue, the embryos would be destroyed. Donors of embryos and adult stem cells should also be informed if their materials will be used to create human–animal chimeras for study and whether they may be likely to be implanted in patients. Participants should also be informed other potential uses of their materials, e.g., to study basic biological mechanisms, to obtain broad genomic data, or test chemicals for toxicity (International Society for Stem Cell Research 2006; National Academy of Sciences 2005; Isasi et al. 2011).

A third approach to informing participants about how their materials will be used is known as tiered approach. The tiered approach gives participants a menu of choices, e.g., check boxes, for how their samples will be used. Participants may grant permission for their samples to be used only in specific types of studies described in the consent document. The advantage of the tiered consent approach is that it promotes autonomy without requiring that participants be recontacted each time that samples are used or shared. A disadvantage of this approach is that it can be difficult for investigators to keep track of and honor the different choices that participants have made, but this problem can be overcome with good record-keeping practices (Salvaterra et al. 2008).

For example, adult stem cell donors might be willing to allow their cells to be used to study basic biological mechanisms and processes or to conduct research on a specific disease, but they might not wish to allow their cells to be used to create chimeras or test toxic chemicals. They might also refuse to allow their cells to be used for organ/tissue transplantation. Embryo donors might refuse to allow their embryos to be used in cloning or in procedures that involve the production of a child. Participants may also want to have the right to decide whether to permit investigators to use samples for commercial purposes or share samples with other researchers (Salvaterra et al. 2008). Participants should also be assured that their materials will not be used in human reproductive cloning. Although the tiered approaches allow investigators to include a check box granting general permission for the use of materials, this option is inadvisable when investigators are procuring embryos for research, due to all the potentially controversial uses of embryos.

7.2.3 *Sources of Materials*

Turning to issues related to the sources of biological materials, no special ethical concerns arise if a competent individual provides adult stem cells. Researchers can obtain consent from that individual. If stem cells are harvested from an aborted fetus, the umbilical cord, or the placenta, then the woman can provide consent. In some instances it is useful to obtain stem cell from children. For example, if a child has leukemia, it may be useful to harvest bone marrow stem cells from the child to study this disease. If a child is a source of the stem cells, consent must be obtained from a parent or guardian. If the stem cells will be kept until the child reaches adulthood (or longer), then the child should be recontacted when he (or she) reaches the age of majority and given an opportunity to decide whether he (or she) wants to continue participating in the research. Investigators should discuss the research with the individual, give him (or her) an opportunity to ask questions, and provide him (or her) with a new consent form to sign. Investigators should explain to the individual how to withdraw samples if he (or she) no longer wishes to participate. As noted above, withdrawing data may not be possible if the data have been used in research, but samples may still be withdrawn. Investigators should not share samples obtained from a minor who has not re-consented upon reaching the age of majority

(Resnik 2009). Although recontacting children who reach the age of majority may present practical difficulties for investigators, it is important to do in order to respect autonomy.

Similar issues concerning consent arise if stem cell or materials are procured from an incompetent adult, such as a person with a mental disability that undermines decision-making. If the person cannot provide valid consent, then consent should be obtained from their legal representative. If the person is expected to be temporarily incompetent, then consent may be obtained from their legal representative and the person should be given an opportunity to decide whether to continue participating in the study if they regain decision-making abilities. If they decide to stop participating in the study, they should be allowed withdraw their samples.

If researchers plan to acquire human embryos, they should obtain consent from both biological parents, unless one is deceased or no longer able to make decisions, to respect their autonomy. If one parent wants to donate embryos for research but the other objects, researchers should not use those embryos. It is important to obtain consent from both parents because they are both impacted by decisions made concerning the use of their embryos (Cohen 2009).

7.2.4 Undue Influence

Another issue related to consent is coercion or undue influence. Coercion involves a threat of harm to a person, while undue influence involves conditions in which the person's ability to make sound decisions may be compromised by financial incentives, social expectations, or other factors (Wertheimer 2010). Research regulations and ethical guidelines require that consent be obtained under conditions that minimize the potential for coercion or undue influence so that participants can make a free (autonomous) choice (Capron 2008). To minimize the potential for coercion or undue influence in the procurement of embryos, many commentators recommend that the decision to create embryos for in vitro fertilization be separated from the decision to donate embryos for research. Couples should be asked to donate leftover embryos only after they have created the embryos for in vitro fertilization and are deciding what to do with the ones they do not use (National Institutes of Health 2009). The NIH guidelines stipulate that the decision to create embryos for in vitro procedures must be separated from the decision to donate embryos for research. This separation helps to ensure that federal funds will not be used to create embryos for research (National Institutes of Health 2009). The individuals who are involved in fertility and reproductive procedures or assist in reproduction and those who request donation of embryos for research should be distinct. The fertility specialist who helps the couples create embryos should not be the same person who procures them for research. This recommendation helps to protect couples from undue influence or coercion from their physician (Cohen 2009). It is important to note, however, that none of these restrictions are necessary if embryos are being intentionally created for research. However, as noted above, the NIH does not fund this type of research.

Procedures to protect individuals from coercion or undue influence can also be implemented in the procurement of adult stem cells from tissues, aborted fetuses, the umbilical cord, or placenta. The individual who treats a patient for a disease or pregnancy and individual who requests a biological sample for research should be distinct, so that the patient will not experience undue influence or coercion from their attending physician. The decision to abort a fetus should be kept separate from the decision to donate fetal tissue for research (Cohen 2007).

Financial incentives to participate in research may constitute an undue influence if they are so high that they interfere with the person's ability to make a sound choice. Some people may so swayed by money that they fail to adequately consider the benefits and risks of participation or weigh their options carefully. They may even lie to investigators in order to qualify for a study (Grady 2005). While many commentators agree that excessive payments for participation may constitute undue influence, establishing an appropriate level of remuneration can be difficult, because many different factors are relevant to compensating participants for their contributions to research, including the degree of pain or discomfort, the amount of time spent participating in a study, and the risks involved. Underpayment can also be problematic because it treats participants unfairly and does not properly acknowledge the value of their contributions (Grady 2005).

The NIH (2009) and many other organizations do not allow researchers to provide payments for embryos, given their special moral status. If one regards the embryo as a human being, then it is unethical to pay for embryos, because human beings have inherent moral worth and should not be treated as commodities. Even if one does not regard the embryo as a full-fledged human being, one could argue that it should still not be treated as a commodity, given its potential to become human (National Academy of Sciences 2005).

Paying research participants for gametes has been controversial. While the selling of sperm is widely accepted due to the negligible risks of this procedure, many are concerned about the selling of ova due to the potential for harm and exploitation (Resnik 2001; Steinbock 2004; Cohen 2007). Egg donation is an invasive and moderately risky procedure that requires the donor to take drugs to stimulate ovulation, which can lead to ovarian hyperstimulation syndrome, a condition marked by abdominal pain and inflammation (Dickenson 2009). Poor women who sell their eggs for money may be especially susceptible to undue inducement and exploitation (Cohen 2007). The National Academy of Sciences (2005) recommends that egg donors receive no payment beyond what is needed to compensate them for their time and inconvenience, and the International Society for Stem Cell Research (2006) recommends that payment not be so high as to constitute undue inducement.

Paying research participants for adult stem cells does not raise the same sorts of ethical concerns as payment for embryos or ova. Most agree that it is acceptable to pay participants for adult stem cells, provided that the payments are not so high as to constitute undue inducement. To decide the appropriate level of payment for participation in a study involving the procurement of adult stem cells, it is important to consider the amount of time required of participants as well as the pain and inconvenience related to study procedures, such as biopsies, blood draws, physical exams, or questionnaires. Physiologic studies involving blood draws, biopsies, physical exams, questionnaires, and other interventions typically pay participants between $50 and $200, though some pay as much as $500 (Grady et al. 2005).

7.3 Confidentiality and Anonymity

7.3.1 Confidentiality

Regulations and ethical guidelines require that investigators protect the confidentiality of human research participants (Department of Health and Human Services 2009; World Medical Association 2008; Hodge and Gostin 2008). Protecting confidentiality in research is important for three reasons. First, participants can be harmed if private information collected during research is disclosed. Participants often provide sensitive information concerning infectious diseases, cancer, substance abuse, or mental illness. A participant could lose his or her job or have difficulty obtaining health insurance if this information is not kept confidential. A participant may also suffer from embarrassment, stigma, or shame. Second, mentally competent participants have a right to control access to their private information, and breaches of confidentiality violate this right. Third, confidentiality protections help to promote participants' trust in researchers and the scientific enterprise (Hodge and Gostin 2008).

As noted above, researchers should inform participants about the steps they are taking to protect confidentiality during the consent process. Researchers can help to protect confidentiality by restricting access to data and samples. Only authorized individuals, such as study staff or collaborators, should have access to data or samples. Data and samples should also be kept in a secure place, such as a locked filing cabinet or refrigerator. Electronic data should be encrypted to enhance security (Hodge and Gostin 2008). Some researchers also apply for a Certificate of Confidentiality (COC) from the Department of Health and Human Services to enhance confidentiality. A COC allows researchers to refuse legal demands to share data, such as requests made in civil litigation. However, COCs may not provide absolute protection against requests to share data, because COCs have not been well-tested in the courts. A court might rule that a COC does not allow investigators to refuse to disclose information necessary for criminal proceedings or national security purposes, for example (Beskow et al. 2008).

The sharing of materials and data in regenerative medicine research raises important confidentiality issues. There are two main ways that regenerative medicine researchers can protect confidentiality when sharing data or materials related to human participants. First, researchers can use data use agreements (DUAs) or material transfer agreements (MTAs) to protect confidentiality. These agreements usually require that recipients not disclose private information about human participants or share it without permission. They also state that researchers may use data or materials only for specified purposes and that they must obtain approval (or an exemption, as may be appropriate) from an IRB or other oversight committee for research conducted with the data or materials. These agreements may also require that recipients not attempt to identify individuals, if they have not been provided with this information (Resnik 2010).

7.3.2 *Anonymity*

Second, researchers can remove information that personally identifies participants (such as a name, address, and social security number) from data or materials when they share it. Removal of identifiers can help protect confidentiality and ensure anonymity of the data. Anonymity is distinct from confidentiality because it may still be possible to identify participants in confidential datasets if identifying information is not removed. There are two types of de-identified data/materials: coded and anonymous. In coded data/materials, researchers assign an alphanumeric code to each participant, but they do not share the key to decipher the code with recipients. The researchers can determine the identity of individuals but recipients cannot. In anonymous data, researchers remove all personal identifiers and do not code it. No one can readily identify individual participants when data and materials are anonymized. To maximize data sharing, some researchers have placed de-identified data on publicly accessible websites, so that scientists can access data readily (Resnik 2010).

These different approaches each have their advantages and disadvantages. Requiring recipients to sign a DUA or MTA maximizes confidentiality protections, but it can also impede research, because it mandates that recipients sign a contract before they can receive data or samples, which takes time and requires legal or administrative support, and may include provisions that limit the use of data or materials. Due to legal liability concerns and intellectual property issues, most institutions require recipients to sign an MTA prior to receiving research materials. Institutions may or may not require recipients to sign a DUA in order to receive data (Shamoo and Resnik 2009).

Removing personal identifiers can help protect confidentiality, but it is not a fail-safe method, since it may still be possible to identify participants. It is often possible to use demographic information, such as zip code, date of birth, and gender, to identify individual participants in de-identified datasets, especially in studies involving small populations (El Emam et al. 2011). Though removal of demographic information can help protect confidentiality, it can also undermine the value of the data and materials, because demographic information is often important for understanding relationships among environmental, social, and racial/ethnic risk factors and disease (Resnik 2010). Bioinformatics methods developed for forensic DNA purposes have made it possible to reidentify de-identified individuals in genomic datasets (McGuire and Gibbs 2006). If one has access to some genetic information that uniquely identifies an individual, then it is possible to find that individual in a database that contains the individual's de-identified genomic information (Homer et al. 2008). Reidentification can also occur when materials are shared, if recipients conduct genetic tests on the materials and attempt to match those results to identified genetic sequences (McGuire and Gibbs 2006). Though some regard the risk of reidentification as remote, others are not willing to take that risk. In response to growing concerns about reidentification, the NIH decided it would no longer place de-identified genomic data from its genome-wide association (GWAS) studies on its public database. Access to GWAS data will only be granted to investigators who sign a DUA (National Institutes of Health 2010, see also Ossorio this volume).

Of special concern to regenerative medicine researchers is traceability, i.e., the ability to identify the donors of specific biospecimens, such as embryos or adult stem cells. Traceability is important for several reasons. First, it is essential to verifying that cell lines meet ethical and legal requirements for informed consent. As noted above, some older HESC lines used in research have not met current consent requirements. Second, traceability is important in case it is necessary to recontact donors to obtain additional clinical or demographic information. Third, traceability gives participants the opportunity to withdraw their samples if they decide to stop participating in research (Isasi et al. 2011).

Since anonymization of samples maximizes confidentiality but undermines traceability, there is a potential conflict between ensuring traceability and safeguarding confidentiality. One could argue, however, that confidentiality can be adequately protected by coding samples, so anonymization is not necessary. Coded samples may be shared, with the investigator retaining the key. If questions arise concerning consent or other issues, then the investigator can be contacted to provide the necessary information.

Because confidentiality is a complex, evolving topic, a case-by-case approach may be the best way of dealing with issues concerning the sharing of data and materials in regenerative medicine. Decisions concerning which type of approach to use should take into account the facts and circumstances. As noted above, most institutions will use MTAs when sharing research materials, but deciding whether to use a DUA for sharing data is an open question. DUAs are appropriate when researchers plan to share identified data or de-identified data in which the risk of reidentification is a significant concern.

7.4 Conclusion

The procurement of human biological materials in regenerative medicine research raises numerous ethical issues, such as informed consent and confidentiality. Though these issues also arise in other types of biomedical research, regenerative medicine investigators face some unique challenges, due to the nature of their subject matter. Since the ethical questions and problems related to procurement of human biological materials are likely to change in response to advances in science and technology, it is important for regenerative medicine researchers to stay abreast of current policies and ethical guidance and to think critically about the dilemmas they face.

Acknowledgements This article is the work product of an employee or group of employees of the National Institute of Environmental Health Sciences (NIEHS), National Institutes of Health (NIH); however, the statements, opinions, or conclusions contained therein do not necessarily represent the statements, opinions, or conclusions of NIEHS, NIH, or the United States government.

References

Aalto-Setälä, K., Conklin, B. R., & Lo, B. (2009). Obtaining consent for future research with induced pluripotent cells: Opportunities and challenges. *PLoS Biology, 7*(2), e42.

Beskow, L. M., Dame, L., & Costello, E. J. (2008). Research ethics. Certificates of confidentiality and compelled disclosure of data. *Science, 322*, 1054–1055.

Brock, D. (2008). Philosophical justifications of informed consent in research. In E. Emanuel, C. Grady, R. Crouch, et al. (Eds.), *The Oxford textbook of clinical research ethics*. New York: Oxford University Press.

Capron, A. M. (2008). Legal and regulatory standards of informed consent in research. In E. Emanuel, C. Grady, R. Crouch, et al. (Eds.), *The Oxford textbook of clinical research ethics*. New York: Oxford University Press.

Cohen, C. B. (2007). *Renewing the stuff of life: Stem cells, ethics, and public policy*. New York: Oxford University Press.

Cohen, C. B. (2009). Ethical and policy issues surrounding the donation of cryopreserved and fresh embryos for human embryonic stem cell research. *Stem Cell Reviews and Reports, 5*, 116–122.

Department of Health and Human Services. (2009). *Protection of human subjects, 45 CFR 46*. Retrieved June 14, 2012, form http://www.hhs.gov/ohrp/humansubjects/guidance/45cfr46.htm.

Dickenson, D. (2009). Good science and good ethics: Why we should discourage payment for eggs for stem cell research. *Nature Reviews Genetics, 10*, 743.

El Emam, K., Arbuckle, L., Koru, G., et al. (2011). The re-identification risk of Canadians from longitudinal demographics. *BMC Medical Information and Decision Making, 11*, 46.

Grady, C. (2005). Payment of clinical research subjects. *Journal of Clinical Investigation, 115*, 1681–1687.

Grady, C., Dickert, N., Jawetz, T., et al. (2005). An analysis of U.S. practices of paying research participants. *Contemporary Clinical Trials, 26*, 365–375.

Hodge, J., & Gostin, L. (2008). Confidentiality. In E. Emanuel, C. Grady, R. Crouch, et al. (Eds.), *The Oxford textbook of clinical research ethics*. New York: Oxford University Press.

Homer, N., Szelinger, S., Redman, M., et al. (2008). Resolving individuals contributing trace amounts of DNA to highly complex mixtures using high-density SNP genotyping microarrays. *PLoS Genetics, 4*(8), e1000167.

International Society for Stem Cell Research. (2006). *Guidelines of the conduct of human embryonic stem cell research*. Retrieved June 14, 2012, from http://www.isscr.org/AM/Template.cfm?Section=GuidelinesforhESCResearch&Template=/CM/ContentDisplay.cfm&ContentID=2916.

Isasi, R., Knoppers, B. M., & Lomax, G. (2011). Sustained interaction: The new normal for stem cell repositories? *Regenerative Medicine, 6*, 783–792.

McGuire, A., & Gibbs, R. (2006). No longer de-identified. *Science, 312*, 370–371.

National Academy of Sciences. (2005). *Guidelines for human embryonic stem cell research*. Washington, DC: National Academies Press.

National Institutes of Health. (2009). *Guidelines for human stem cell research*. Retrieved May, 29, 2012, from http://stemcells.nih.gov/policy/2009guidelines.htm.

National Institutes of Health. (2010). *Policy for sharing of data obtained in NIH supported or conducted genome-wide association studies (GWAS)*. Retrieved June 14, 2012, from http://grants.nih.gov/grants/guide/notice-files/NOT-OD-07-088.html.

Petrini, C. (2010). "Broad" consent, exceptions to consent and the question of using biological samples for research purposes different from the initial collection purpose. *Social Science and Medicine, 70*, 217–220.

Resnik, D. B. (2001). Regulating the market for human eggs. *Bioethics, 15*, 1–25.

Resnik, D. B. (2004). *Owning the genome: A moral analysis of DNA patents*. Albany, NY: State University of New York Press.

Resnik, D. B. (2009). Re-consenting human subjects: Ethical, legal and practical issues. *Journal of Medical Ethics, 35*, 656–657.

Resnik, D. B. (2010). Genomic research data: Open vs. restricted access. *IRB, 32*(1), 1–6.

Resnik, D. B. (2011). Disclosure of individualized research results: A precautionary approach. *Accountability in Research, 18*, 382–397.

Salvaterra, E., Lecchi, L., Giovanelli, S., et al. (2008). Banking together: A unified model of informed consent for biobanking. *EMBO Reports, 9*, 307–313.

Shamoo, A. S., & Resnik, D. B. (2009). *Responsible conduct of research* (2nd ed.). New York: Oxford University Press.

Skloot, R. (2010). *The immortal life of Henrietta Lacks*. New York: Crown.

Steinbock, B. (2004). Payment for egg donation and surrogacy. *Mount Sinai Journal of Medicine, 71*, 255–265.

Streiffer, R. (2008). Informed consent and federal funding for stem cell research. *Hastings Center Report, 38*(3), 40–47.

Weir, R., & Olick, R. (2004). *The stored tissue issue*. New York: Oxford University Press.

Wendler, D. (2006). One-time general consent for research on biological samples. *British Medical Journal, 332*, 544–547.

Wertheimer, A. (2010). *Rethinking the ethics of clinical research: Widening the lens*. New York: Oxford University Press.

Wolf, S. M., Lawrenz, F. P., Nelson, C. A., et al. (2008). Managing incidental findings in human subjects research: Analysis and recommendations. *Journal of Law, Medicine & Ethics, 36*, 219–248.

World Medical Association. (2008). *Declaration of Helsinki*. Retrieved June 14, 2012, from http://www.wma.net/en/30publications/10policies/b3/index.html.

Chapter 8
Early-Stage Research: Issues in Design and Ethics

Nancy M.P. King

8.1 Introduction

Most regenerative medicine research is still in the earliest stages; few interventions have as yet moved from laboratory and animal studies into humans. It could be argued, then, that human research ethics considerations are largely premature in regenerative medicine. However, the importance of ethical reflection to translational research is highlighted by a growing literature and lively scholarly discussion of "first-in-human" or "bench-to-bedside" clinical trials. It is therefore not too early to consider both the science and the ethics of translational research in complex and promising novel biotechnologies like regenerative medicine.

This chapter examines ethical issues in "early-stage" regenerative medicine research. Early-stage research is an imprecise term that includes not only classic phase I and phase II studies but also novel and hybrid study designs, all corresponding roughly to research that precedes the large studies that are designed to definitively demonstrate the clinical utility of a treatment candidate. Specific attention to the ethical and policy issues arising in early-stage research is particularly important in regenerative medicine. This broad and promising field of interdisciplinary science is having a profound effect on how bench-to-bedside research is understood and conducted, both because it combines research perspectives from diverse fields and because it is one of the novel biotechnologies that are helping to shape a more dynamic understanding of bench-to-bedside translation.

This chapter addresses several of the issues most relevant for regenerative medicine research. It begins with the term "regenerative medicine" itself and acknowledges the breadth and complexity of the field. Next it considers the ethical and policy implications of the timing of the translational trajectory and the difficult

N.M.P. King, J.D. (✉)
Department of Social Sciences & Health Policy and Wake Forest Institute for Regenerative Medicine, Wake Forest School of Medicine, Wake Forest University
e-mail: nmpking@wfubmc.edu

L.F. Hogle (ed.), *Regenerative Medicine Ethics: Governing Research and Knowledge Practices*, DOI 10.1007/978-1-4614-9062-3_8,

balance between learning and treating; how should regenerative medicine researchers determine when to move into humans? Finally it examines how the therapeutic misconception, that is, misunderstanding research as treatment, can affect potential subjects, researchers, oversight bodies, and media reporting in regenerative medicine research.

8.2 "Regenerative Medicine:" A Field and Its Terms

The term "regenerative medicine" describes an extraordinarily broad and diverse realm of research, unified by the scientific quest to understand the basic mechanisms of generation, or the growth and development of living organisms (King et al. 2010). The development and application of that understanding takes a wide range of forms, ranging from in situ cell-based interventions to ex vivo organogenesis using novel combinations of tissue engineering and biomaterials (Atala et al. 2010; King et al. 2010). It is worth noting, however, that the umbrella term "regenerative medicine," as useful as it is, may be perceived very differently by the general public from the way it is perceived by the scientific community. A similar concern arises with the terminology used for subfields within regenerative medicine, such as "organ regeneration" and "cell therapies." Each of these terms is intended to describe research interventions but also carries the implicit connotation of standard, approved, successful treatment, and thus presents at least some risk of inducing or exacerbating confusion. The therapeutic misconception is discussed in greater depth later in this chapter; for the present, two observations can be made. First, the history of gene transfer research, called "gene therapy" from the time of the first human study in 1990, provides a clear parallel. Even now, just a handful of early-stage gene transfer trials have shown promise; yet therapeutic terminology has been applied to the field since its beginnings and has influenced both media discussions and public perceptions about the current state and promise of that research (Churchill et al. 1998). Second, although "cell therapy" has become the standard term to describe that large part of regenerative medicine research that studies and makes use of human embryonic stem cells, hematopoetic stem cells, induced pluripotent stem cells, and various other categories of multipotent and pluripotent stem cells, there is an alternative and better term. Dr. Jeremy Sugarman's "cell-based interventions" avoids unnecessary and misleading therapeutic connotations and focuses properly on the cells themselves, by using the neutral term "intervention."

8.3 Moving From Bench to Bedside

Every investigator doing preclinical research in an area of translational science has reason to think long and hard about the research trajectory. Even though accurate prediction of where a new line of scientific inquiry may lead is difficult, even impossible, it is nonetheless essential to consider carefully and revisit repeatedly questions like: What is the long-term goal of this line of research? What will it take to get there, and what data will be needed to demonstrate whether we have arrived? The design of preclinical translational research in regenerative medicine is complex, situation-specific, and governed by regulatory requirements that are often challenging to follow, but guidance improves as progress is made (see, e.g., Frey-Vasconcells et al. 2012). Most pertinent for this chapter: When is the right time to move from laboratory and animal studies into humans?

This last and profoundly important question of design and ethics requires a complicated balancing between the perspectives of basic science and clinical medicine: "Basic scientists want to know; clinicians want it to work" (King and Cohen-Haguenauer 2008 at 433). When the potential to decrease human suffering is on the horizon, the desire to move safely but rapidly into humans is understandably strong. Moreover, pressures for commercialization and product development may also increase the expected speed of bench to bedside translation. Each experimental step in any research trajectory must of course be designed so that the data gathered are likely to reduce the uncertainties associated with the intervention under study; it is only in this way that a line of research moves forward. However, the move from preclinical to clinical trials is scientifically and ethically sound only when three additional conditions are met:

1. No more can reasonably be learned without delivering the experimental intervention to human research subjects.
2. The risks of harm and amount of uncertainty have been reduced as far as feasible.
3. The remaining risks of harm and uncertainties are not excessive under the circumstances.

Note that the operative terms here are imprecise: reasonable, feasible, not excessive. Precision with regard to these evaluative terms is unobtainable; instead, the proper goal is thoughtful and productive examination and discussion of their applicability to the unique circumstances of a particular line of research.

How should regenerative medicine researchers go about deciding whether an intervention in the preclinical stages is ready for the "prime time" of human subjects research? The evaluative criteria listed above will apply differently to every line of research, but the Geron Corporation trial of a human embryonic stem cell-derived intervention for subacute thoracic spinal cord injury may provide a helpful example.

Early in 2009, the Food and Drug Administration (FDA) approved Geron's trial for spinal cord injury, which was the very first clinical trial using human embryonic stem cell-derived tissues (Bretzner et al. 2011). This phase I study, like all phase I studies, was focused first and foremost on determining safety. However, as is the case with many phase I trials, only patients were to be enrolled as subjects, in part because the safety profile would be different in healthy volunteers and in part because if any preliminary efficacy signals could be seen, they would of course only be seen in patient-subjects. These subjects were patients who had experienced very serious spinal injury very recently, making them different from many patient-subjects in phase I studies, who most often are those for whom all other treatments have failed (King et al. 2009 at 225 and note 51).

Research ethics questions arose about the trial's design because preclinical studies were apparently limited to small animal models, and because the first subjects were those with a very serious, very recent injury and a small post-injury enrollment window, leading to concerns about the amount and quality of information available for informed decisionmaking in a short time (Bretzner et al. 2011). The study was placed on clinical hold in August 2009, after ongoing review of preclinical animal data raised the possibility of tumorigenesis and thus the need for additional purification of the cell lines used (Bretzner et al. 2011). The clinical hold was lifted a year later, and the trial resumed. However, in November of 2011, after four of the planned 8–10 patient-subjects had begun trial participation, Geron announced that it was ending the trial and shifting its business emphasis away from embryonic stem cell research to focus on pharmaceuticals farther along in the trial pipeline (Ichim et al. 2011; Frantz 2012). Geron's decision suggests the possibility that the company may have moved prematurely into its phase I clinical trial and underestimated the time and expense necessary to complete the research trajectory (Ichim et al. 2011).

8.4 Validity in Regenerative Medicine Research

Once the decision has been made to begin human trials, ethical issues emerge more clearly alongside scientific questions in research design and conduct. One important issue immediately arises regarding the trajectory of translational research: How should the human research trajectory be designed and understood?

The classic phases of research with human subjects (phases I, II, and III) are clearly defined and specified by regulatory and oversight bodies and a great deal of scholarly literature and scientific guidance. However, the reality of clinical research has always been far more fluid than these phase designations, which were developed originally for pharmaceutical research. In particular, when experimental interventions moved from pharmaceuticals to biotechnologies, and when surgery began to be widely and systematically viewed as subject to the requirements of research, these phase designations began to be recognized as often fitting poorly with the realities of research design and subject selection. Regardless, every clinical trial must meet the basic requirements of value and validity (Emanuel et al. 2000): Every trial must ask a scientifically and socially meaningful question and must be designed so that the data gathered have the capacity to answer that question, whether positively or negatively. Perhaps most significantly, the answer to the question must be able to point to the appropriate next step in the research trajectory. In classic phase design, the results of a phase I study should facilitate the move to a phase II study or signal the need for a return to preclinical research or a differently designed phase I study. Even when the clarity of phase designations has been blurred by novel designs and trajectories, every trial should produce results that are valid—that is, able to guide researchers in designing the next study, whatever it may be (Kimmelman 2009).

For complex and varied fields like regenerative medicine, the requirement of validity may have some unfamiliar implications. Regenerative medicine research may be complex and multidisciplinary, involving surgical innovation, tissue engineering, and cell-based interventions. An expectation of benefit may be incorporated into even a first-in-human trial. Moreover, the risks of harm arising from failure may be potentially catastrophic, in particular because it may not be possible to "undo" a complex biological intervention in the same way that a drug can be discontinued in a pharmaceutical trial.

A complicated regenerative medicine intervention may take considerable time, and follow-up can easily be lifelong. Phase designations may not make logical sense for this type of research, especially because the surgeons and scientists who perform it are far more likely to employ revised, refined, and improved processes, materials, and techniques for every enrolled patient-subject at the next stage of an ongoing procedure, or to refine their techniques with later-enrolling patient-subjects if the intervention is more limited. Those who enroll early may not be in a position to benefit from all later-developed improvements. Thus, the realities of regenerative medicine research may be far more fluid than is commonly seen in many translational trajectories.

8.5 Ethical Issues in Early-Stage Research: Two Examples

Two examples drawn from the many types of regenerative medicine research may help to illustrate some of the principal ethical issues arising in early-stage regenerative medicine research. The first example is current and real; the second represents an important area of research concentration that has not yet moved beyond the basic laboratory stages. The two examples are set forth below, so that readers can bear them in mind as key ethical issues in early-stage regenerative medicine are discussed. The examples are then revisited at the end of the chapter and linked to the ethical issues that have been examined.

8.5.1 Bladder Augmentation (The Real Example)

Dr. Tony Atala's groundbreaking research uses determined (or "adult") stem cells from the bladders of patient-subjects with meningomyelocele to grow a larger segment of that hollow organ. The new segment is then surgically attached to the existing organ, in order to increase bladder capacity in patients whose condition has prevented ordinary bladder growth. The goal of bladder augmentation for patients with meningomyelocele is to make it possible for them to self-catheterize less frequently, thus improving quality of life and reducing the likelihood of kidney damage from urinary back-pressure, but bladder augmentation cannot alter the absence of contractility resulting from the spinal lesion. Bladder augmentation using segments of intestine has been surgically tried and has become standard of care, but it is not always successful and entails considerable postoperative morbidity.

In Atala's study, autologous determined urothelial and smooth muscle stem cells were harvested from pediatric patient-subjects, cultured in vitro, grown on decellularized scaffolds, and then surgically implanted. Patient-subjects in this small study were followed for approximately 5 years before the investigators were confident that this first-in-humans intervention could be deemed successful and its results published (Atala et al. 2006).

8.5.2 Limb Regeneration (An Imagined Future Example)

The "human salamander" represents a Holy Grail of sorts for regenerative medicine. The ability to regenerate entire limbs exists in the animal kingdom, and the desire to unlock the secrets of total regeneration is closely linked to the recognition that this knowledge could provide important basic understanding of the mechanisms of both aging and cancer. More immediately, limb regeneration research seeks the means to ameliorate traumatic injuries, especially those suffered in combat by soldiers and civilians. Some research has demonstrated regeneration of minor injuries (e.g., the tip of a finger), but regeneration of a complete extremity is as yet a distant hope. Imagine, however, that regeneration of a forearm amputated below the elbow is thought to be a realistic possibility, based on laboratory studies of in vitro limb culturing and bioreactor development (much like the work that has been done with ears) and on a limited number of animal studies (perhaps mice receiving research amputations of forelimbs, and a small study of several nonhuman primates with accidental injuries, or birth deformities such as amniotic band syndrome).

8.6 Subject Selection

Once a line of research is considered ready to take into human trials, a key decision must be made: Who should be the first research subjects? At every stage of clinical research, the choice of subjects must balance two criteria that may be at odds: First, subjects must be those who can provide the best data; that is, data that will enable determination of the appropriate next research step, whether forward, back to gather additional data, or sideways to pursue a different path (Kimmelman 2009; King and Cohen-Haguenauer 2008). Second, at the same time, subjects must also be at the least risk of harm to themselves from research participation. Harm here includes both harm arising directly from receiving the research intervention and the harm that may arise if any standard interventions must be detrimentally forgone or postponed in order to participate in the research (Kimmelman 2005; King and Cohen-Haguenauer 2008).

Many first-in-human safety trials of pharmaceuticals have conventionally enrolled healthy volunteers, but it has become common, in many areas of research, to enroll as first subjects patients who have the disease or condition of interest.

Healthy volunteers may not always provide the best data under the circumstances, as they are often poor models for the disease states under study. In addition, many interventions (e.g., chemotherapy, surgery, and novel biotechnological interventions like gene transfer) may pose significant risks of harm to subjects, and it is widely regarded as unethical to ask healthy volunteers to submit to invasive research interventions bearing significant risks of harm. Thus it is often reasonable to turn first to patients as research subjects.

An unintended consequence may unfortunately result: the mistaken belief that, if healthy volunteers are not enrolled, it is because they cannot benefit, but because patients will be enrolled, they stand at least some chance to benefit, almost by definition, even in early-stage research. This belief is one manifestation of the therapeutic misconception, discussed later in the chapter.

Even after patients are chosen as subjects, investigators must consider which among them should be first. Patients who are younger and/or whose disease has not progressed very far often represent a very different population from older patients. The same is true of so-called treatment-naive patients in comparison with those who (as in most oncology research) have tried many treatments and can derive no further benefit from standard therapies.

8.7 Harms and Benefits, Uncertainties and Unknowns

8.7.1 Balancing Risks of Harm and Potential Benefits

Investigators and research oversight bodies face a difficult but essential challenge in determining whether a first-in-humans trial may justifiably move forward: They must agree that the potential benefits to society from the likely results of the trial and the potential benefits to individual patient-subjects in the trial—if any are possible, of course—outweigh the risks of harm to patient-subjects in the trial (King and Churchill 2008). The assessment and balancing of benefits and harms is especially challenging in the context of early-stage clinical research, when what is uncertain and what is unknown loom particularly large. This balancing should be the subject of genuine conversation between investigators and oversight bodies, and it must precede determination of what information should be provided to potential subjects in the consent form and process. Yet asking the question "How should this research be described to potential subjects?" can fruitfully inform that logically prior discussion, clarifying precisely what risks of harm and potential benefits are at issue, and helping to determine whether offering research participation to potential subjects is justifiable under the circumstances.

Because early-stage research in regenerative medicine comprises such a broad category of research types, from stem cell-based interventions to complex tissue engineering experiments, it is extraordinarily difficult to generalize about risks of harm, about the scope of potentially relevant uncertainties and unknowns, about benefits reasonably to be expected in a given study, and even about how harms and benefits should be identified and measured. The following brief discussions represent only the tip of the translational iceberg; our two examples provide some additional food for thought; but the ultimate goal of this chapter is to equip readers to apply and extend these points to consider when planning and carrying out specific translational studies.

8.7.2 *Risks of Harm*

Disclosure and description of the risks of harm—from receipt of the experimental intervention, from research procedures employed adjunctively for monitoring and follow-up, and, in some studies, from forgoing available alternatives—is a well-rehearsed area of research ethics (Kimmelman 2004) that will not be exhaustively detailed here. Just a few points are especially relevant and important in early-stage regenerative medicine research.

As in all early-stage research, all anticipated risks of harm must be described thoroughly and clearly, enumerating what is known or expected about their nature, magnitude (size/intensity and duration), and likelihood. Then, in addition to describing and discussing the risks of harm from the experimental intervention itself, it is also important to inform potential subjects that whenever alternatives exist (such as standard treatments, however imperfect), harm could result from forgoing or delaying standard treatment. Finally, open acknowledgment of the possibility of unknown effects is essential.

8.7.3 *Potential for Benefit*

Determining whether any potential benefits may reasonably be expected from the experimental intervention, and therefore should be disclosed and described to potential subjects, is a pervasive and significant challenge, both for research oversight and for informed consent in human subjects research. This determination is extremely important and especially difficult in early-stage research, where direct benefit is often unlikely or impossible (King 2000). Clarity about potential benefit can easily be outweighed by excessive expectations, especially under circumstances that are increasingly common in early-stage research involving complex biotechnologies like regenerative medicine. In early-stage research, available information about potential benefit is limited at best; the goal of the line of research is to demonstrate clinical benefit; and the potential subjects in early-stage trials are patients with the disease or condition that the experimental intervention is ultimately intended to treat. Thus, excessive expectations may affect not only potential subjects, but even more importantly, clinical investigators, members of research oversight bodies, and even study sponsors may also have high hopes that affect how information about an early-phase clinical trial is shaped and shared.

Unfortunately, the discussion of potential benefit in research consent forms—particularly consent forms in early-stage research—is frequently vague, stereotypical, uninformative at best, and misleading at worst (King et al. 2005). Careful attention to describing and discussing potential benefits, and their limits, can be especially challenging for researchers, because they often receive little help from funders and research oversight bodies. The model consent forms and guidance provided may even exacerbate the difficulty. Defining and describing potential benefit well is a problem that merits more attention in early-stage research and should be addressed by funders and oversight bodies as well as by researchers.

In every clinical trial, and thus in all regenerative medicine research, the potential for direct benefit is intervention-specific. The potential benefits of different types of regenerative medicine interventions may be quite different in nature, magnitude, and likelihood, even though the effects of at least some interventions may be pervasive or permanent. And it is undeniable that some effects of regenerative medicine interventions are unanticipated and unknown (Kimmelman and London 2011).

Anticipated benefits for a particular early-phase clinical trial should be thoroughly and clearly described, and the basis for any expectations should be explained. The potential for benefit may in some instances be nonexistent; expectations of benefit may be theoretical only; or benefit may be possible but unlikely, based on preclinical evidence from laboratory and animal models, or analogized from similar but not identical research. Thus, it is important to emphasize the limits on what may be anticipated in a given trial.

Several components of the discussion of potential benefit are particularly important in early-stage research, including regenerative medicine research. First, it is essential to distinguish direct, inclusion, and societal benefits. Direct benefits are those arising from the experimental intervention. These benefits are likely to be of greatest significance for patient-subjects. Inclusion benefits arise from simply participating in the research, whether or not one receives the experimental intervention or is benefited by it. Inclusion benefits are provided to all subjects and are sometimes conceptualized or described as inducements to participate. Examples include a free physical examination, medical testing and monitoring beyond what is required by the research, and other nonmonetary benefits. Sometimes, the additional testing and monitoring required by the research is described as beneficial: "Patients do better on study because we monitor them so closely." It is worthy of note that close monitoring and additional testing may just as easily pose risks of harm, as when research-related testing gives a "false positive" result that produces anxiety and requires additional testing at the patient-subject's expense. Finally, societal benefits stem not from research participation but from the outcomes of the line of research (King 2000; King et al. 2005).

Thus, both in consent forms and in discussion with research oversight bodies, it is necessary to use language about potential benefit very carefully and deliberately. To give a simple example: "You may benefit from being in this study" is different from "You may benefit from getting the experimental intervention." The former refers to inclusion benefits, and the latter to direct benefits. Similarly, "The purpose of this study is to find out whether or not subjects can benefit from getting the experimental intervention" is different from "The purpose of this research is to develop a new treatment for X disease." The former is specific to the study at hand, while the latter confuses the line of research with the study at hand and may suggest that treatment benefits are anticipated for patient-subjects (King 2000; National Institutes of Health nd).

Second, discussion of the potential for direct benefit must be more specific and detailed than the all-too-common boilerplate statement, "You may or may not benefit." Direct benefit can and should be described in terms that resemble the more familiar description of risks of harm: the nature of the benefit, its magnitude (i.e., its size and duration—a change in laboratory values, which may or may not be linkable to clinical benefit? a reduction of symptoms? a cure? a temporary or permanent effect?), and its likelihood. In early-stage research, especially research with levels of uncertainty as high as in most regenerative medicine research, these dimensions of direct benefit may be exceedingly difficult to quantify; nonetheless, addressing them, even when precision is impossible, at least signals to potential subjects that there is more to the potential for benefit than "Either I will benefit or I won't" (King et al. 2005).

One of the challenges in discussing potential benefit, which is especially acute in all early-stage research, is avoiding overdisclosure in the consent form and process. Clear and thorough description may become so detailed that not only potential subjects but also oversight bodies and investigators can fall prey to "information seduction"—whereby the sheer amount of information provides a false assurance that clinical benefit will materialize. Underdisclosure is not a solution to this problem; it merely perpetuates uninformative "boilerplate." Instead, open acknowledgment of uncertainty should be coupled with the promise to minimize risks of harm, whether or not potential benefit is anticipated, disclosed, and described.

8.7.4 Uncertainties and Unknowns

Early-stage trials present far more uncertainty than later-phase studies. A primary reason is that the translation from preclinical to clinical research represents a very large step. Good animal models are not always available, and available models are always imperfect. Thus, early-stage trials often begin with high levels of uncertainty about the effects of the intervention in humans. The wide variety of experimental interventions in regenerative medicine—cell infusions of many types, surgical implantation of cell-based materials from wafers to organs, in situ regeneration, and more—all carry many (and many different) uncertainties and all can change as rapidly as the science develops.

This lack of knowledge poses challenges for informed consent, particularly with regard to how best to describe and discuss what is uncertain and what is unknown. This is a problem of disclosure, to be sure; but it also is a key factor in the consideration of whether a study is ready to move from preclinical to clinical trials (King and Cohen-Haguenauer 2008). How best to determine this critical readiness remains underaddressed, especially for novel biotechnologies like regenerative medicine interventions (Frey-Vasconcells et al. 2012).

8.7.5 Long-Term Follow-up

Understanding the role of patient-subject includes understanding why long-term follow-up is sought—both to gather optimal study data and to properly protect the welfare and interests of patient-subjects. But when subjects are lost to follow-up, most of the problem may not lie with the consent form and process. Instead, the design, funding, and infrastructure needed for good long-term follow-up may be inadequate to sustain it. Thus, investigators need to design good long-term follow-up and incorporate it into their protocols, and funding agencies need to support recommended follow-up. Investigators also need to address the practicalities of long-term follow-up. Making it easier for patient-subjects helps underscore its importance and their role in knowledge production.

Determining what follow-up is necessary, appropriate, and practical is a study-specific exercise, based on the nature of the regenerative medicine intervention being studied and the information being sought. Long-term follow-up in regenerative medicine research may be necessary over many years simply in order to determine whether an experimental intervention is a success or a failure. Cell-based interventions may need long-term follow-up so that investigators can learn about patterns of bioaccumulation and their potential effects, such as insertional mutagenesis, as has been seen in some gene transfer research, or other tumorigenic effects, as may be seen when embryonic or induced pluripotent stem cells are used. Long-term follow-up is also likely to be extensive in any study involving the implantation of organs or tissues regenerated ex vivo, to ensure that success or failure can be determined in functional terms.

Finally, some categories of regenerative medicine research, such as in situ regeneration of organs, digits, or limbs may present a special set of issues. In these types of research, long-term follow-up may be especially extensive, for two reasons. First, in situ regeneration is likely to take some time. Second, in early-stage research, it will not be known right away how long successful regeneration will take, so that long-term follow-up will help to determine both whether the experimental intervention is "working" and how long monitoring should last before success or failure is declared. Significant differences in follow-up may be anticipated, depending on the intervention being monitored. In situ regeneration of a solid organ, such as a kidney or pancreas, may, for instance, be determined successful when a minimum level of organ function has been reached, so that dialysis or insulin is not necessary. Measuring success in the regeneration of a finger or hand may take far longer, and whether a partial regeneration can be deemed successful is far less clear.

For these reasons, it will be wise to ensure that patient-subjects are given ample information about what to expect from any regenerative medicine intervention and the anticipated challenges of long-term follow-up. In addition, it seems prudent to discuss these matters from the particular perspectives of individual potential subjects, as different individuals could reasonably hold very different views about what counts as success or failure; moreover, those views could reasonably change over the duration of study participation and follow-up. Periodic revisiting of discussion about research participation and long-term follow-up should ideally be an integral part of an ongoing researcher–subject relationship (National Institutes of Health nd).

8.8 The Therapeutic Misconception: Causes and Cures

The therapeutic misconception, widely identified and discussed in clinical research enrolling patients as subjects, is the tendency to view research as treatment, to blur the distinction between research and treatment, and/or to have unreasonably high expectations of direct benefit from receiving the experimental intervention. First identified by Paul Appelbaum and colleagues some 30 years ago (Appelbaum et al. 1982), the therapeutic misconception is most often attributed to patient-subjects, but it is vital to recognize that it is also common in investigators and oversight bodies (Dresser 2002). The therapeutic misconception is of concern because it may adversely affect understanding about the nature of the research and the likelihood that the experimental intervention will be beneficial for subjects. It thus might, but does not necessarily, compromise decisionmaking by patient-subjects. More importantly, it might also influence how investigators describe the research to potential subjects in the informed consent process, as well as how oversight bodies like institutional review boards (IRBs) view the research (Churchill et al. 2003; Henderson et al. 2006; Miller 2000). A great deal of scholarly literature has dissected the concept of the therapeutic misconception, but close examination of that literature is beyond the scope of this chapter; the term "therapeutic misconception" is therefore employed throughout as a well-recognized idea, even though more nuanced terminology might appear in a more extended discussion.

Although the therapeutic misconception may influence decisionmaking in any clinical trial enrolling patients as subjects, it may be more likely in early-stage research, for several reasons. First, because in most early-stage trials, all patient-subjects receive the experimental intervention, early-stage research designs may be more likely to seem "treatment-like" to all concerned than do randomized, placebo-controlled trials. Second, the patients approached for participation in early-stage trials are often those with severe or advanced disease, for whom there are no good treatment choices available, either because all standard treatments have failed or because no good standard treatment exists (Dresser 2009). This is the approach to subject selection that is most often employed in oncology research (King and Cohen-Haguenauer 2008; King et al. 2005). In such cases, all involved—investigators and IRB members perhaps even more than patient-subjects—are hoping that a new, untried intervention will offer some benefit that standard treatment cannot provide (Dresser 2002). Although this hope is understandable, the resulting therapeutic misconception may have significant distorting effects on decisionmaking.

It is not yet clear how best to identify the therapeutic misconception and assess its effects on decisionmaking in clinical research (Henderson et al. 2007). However, the likelihood of the therapeutic misconception in patient-subjects enrolled in early-stage research can be considerably reduced if it is addressed and reduced in investigators and IRB members, so that the consent form and process provide clear, accurate, and realistic information about the potential for direct benefit (Henderson et al. 2006; King et al. 2005). Second, the possibility of the therapeutic misconception should never automatically disqualify patients as potential subjects, especially when vague or misleading information about potential benefit has contributed to their views. Hope for benefit is not always therapeutic misconception; it is acceptable, even desirable to hope for benefit if you don't expect it. Even a subject who says, "I know that the likelihood that anyone in this trial will experience any meaningful benefit is 1 in 100; but I'm confident that I will be that one!" may be expressing a degree of optimism that is unproblematic in context (Sulmasy et al. 2010; Horng and Grady 2003).

Regenerative medicine research may, generally speaking, be especially vulnerable to the therapeutic misconception for several reasons. First, public confusion abounds about the different types of stem cells used in treatment and research (King et al. 2011; Ichim et al. 2011). Because there are many therapeutic uses for determined (adult) stem cells, both autologous and allogeneic, it is not surprising that these long-standing therapies might be confused with research uses of multipotent and pluripotent stem cells. As noted earlier, the very term "cell therapy" invites such confusion. Second, the burgeoning market for stem cell tourism involves use of various types of stem cells in therapies that are offered without having first been tested in research (Cohen and Cohen 2010a, b). Third, the development of for-profit autologous banking of umbilical cord blood stem cells, amniotic fluid stem cells, and even determined stem cells from body fat has helped convey the impression that all cell therapies are truly therapeutic.

A fourth factor that puts other forms of regenerative medicine at risk for therapeutic misconception is that many interventions have a surgical component. Surgery, like chemotherapy, is generally not practiced on healthy volunteers, and a surgical intervention is difficult to limit to safety considerations in an early-stage trial (King 2003). (Moreover, randomized trials are unlikely to be feasible even at later stages of regenerative medicine research, with the possible exception of some comparative effectiveness trials.) Thus, hopes of benefit are likely to exist for all involved in early-stage regenerative medicine research of many types. To reduce the therapeutic misconception, those hopes must be acknowledged and directly addressed.

8.9 Two Examples: Following Up

8.9.1 Bladder Augmentation

This early-stage regenerative medicine trial enrolled children with meningomyelocele, rather than adults, as the most appropriate patient-subjects because they had not yet suffered irreversible kidney damage as a complication of their condition.

These patient-subjects had exhausted all nonsurgical alternatives and were candidates for other types of surgical augmentation of their bladders. The use of a segment of intestine is the most common standard alternative, but it has had limited success. Enrolling in the trial merely delayed but did not foreclose later use of standard surgical alternatives.

Notably, the design of the study changed over time, so that even in a trial with a small number of subjects, techniques evolved to increase effectiveness and were tried in later subjects. This serves as a reminder of the degree of uncertainty that accompanies research and of the goal of research: generalizable knowledge for the benefit of future patients.

Continuing follow-up for 5 years before declaring the trial successful may seem to some overly conservative, but Dr. Atala's decision reflected consideration of the many parameters that were followed and described in order to illustrate the criteria for success. Interestingly, media coverage of organ regeneration research at the Wake Forest Institute for Regenerative Medicine several years later demonstrated the ease with which early-stage research can be mischaracterized and misunderstood by the media and the lay public, who are also susceptible to the therapeutic misconception. In 2011, Dr. Atala gave a TED talk. He briefly described preliminary work in organ regeneration, including the use of bioprinting devices to build collagen scaffolds on which solid organs might be "seeded" using determined stem cells derived from individual patient-subjects—much like what had been done in the bladder augmentation study, but using as an example a bioprinted matrix shaped like a miniature kidney. In addition to displaying the bioprinted kidney matrix, which he described as a nonfunctioning model, Dr. Atala was reintroduced to one of the patient-subjects from the bladder augmentation trial, now a young adult, who noted that his kidneys would have failed if he had not participated in the trial (Atala 2011).

A freelance reporter at the TED conference apparently misunderstood the presentation and stated on a web-based news site that Dr. Atala had bioprinted a functional kidney. This incorrect and scientifically unverifiable account was widely picked up and re-reported by other web-based news outlets. The error was corrected, but only after much effort by Institute media staff. This example demonstrates how easy it is to slip into the therapeutic misconception—and how our faith in scientific progress makes everyone in society susceptible to it.

8.9.2 Limb Regeneration

Basic research into the mechanisms of limb regeneration has been underway for some years, funded in part by the Department of Defense, with the goal of improving treatment for injuries suffered from improvised explosive devices by military personnel and civilians. When research has progressed sufficiently, the determination must be made whether the harm-benefit balance supports moving to human trials. Should mammalian models of limb regeneration be developed, or will it be ethically and scientifically appropriate to move directly into human subjects? Can alternatives to animal models (computer modeling, in vitro and bioreactor work, etc.) be substituted? The answers to these questions will no doubt give rise to a number of different research trajectories, which could vary by many research dollars and many years in length.

When the decision to move into humans has been made, the first patients likely to be proposed as subjects for limb regeneration studies of many types—ranging from partial regeneration to improve an artificial limb interface to complete regeneration—will be casualties of war and others who have been recently and traumatically injured. Important questions will arise about the so-called decision-making window—will potential subjects have to decide soon after sustaining a serious injury whether they wish to become research subjects or to receive the standard treatment (potential amputation and future fitting with artificial limbs)? Or will the window be bigger? Will it be possible to seek standard treatment first, and enroll in a limb regeneration trial later, if the artificial limb options are unsatisfactory? What about choosing research participation first, and switching to standard treatment later, if the research is not successful?

That is, will a choice between research and treatment be reasonably possible for a person in the potential subject's position? Will the timing and context of the decision require special attention to promote autonomous and informed choice by potential subjects in stressful circumstances? When choices by potential subjects are reasonably possible, will choosing between research and standard treatment require divergence between two paths that cannot later reconverge?

Path divergence may indeed be likely for those who choose research participation first. Our example, regeneration of a forearm and hand, will probably be a lengthy process. It may take years before it can fully be known whether the process is complete enough for the regenerated limb to be sufficiently functional. Intensive long-term follow-up is likely to be necessary in order to learn how best to promote rapid regeneration without adverse effects (e.g., oncogenesis), and modifications in the intervention are more likely to be effectively applied to later subjects than to those from whom they are learned. Thus, the patient-subjects in this early-stage research will be asked to make a considerable commitment to the research path, without changing paths to pursue standard treatments if they get tired of waiting for functional regeneration. They will have important roles in defining and refining functionality. They cannot be precluded from dropping out of the research if they are unsatisfied. But they may not get what they expect—or what they really want.

8.10 Conclusion

The examples of bladder augmentation and limb regeneration reflect only a very small portion of the wide variety of regenerative medicine research. Nonetheless, they help illustrate the most common ethical issues arising in the challenging realm of early-stage regenerative medicine research: when to move to human trials, whom to enroll first, how best to inform them, how best to protect them, and how to gather the best data at the least human cost.

Those who undertake early-stage regenerative medicine research should be expected, by funding, review, and oversight bodies, potential subjects, and the public, to explain and justify the scientific need for and ethical appropriateness of:

1. Moving from preclinical research to human subjects.
2. The group of subjects chosen for the particular study.
3. The design features chosen to minimize harms and lost opportunities for those subjects and to maximize the value of the data collected in order best to determine the next steps in the translational research trajectory.

The task of the clinical researcher, in essence, is to make a fair offer of research participation to potential subjects under conditions of uncertainty, where the goals are twofold: to contribute to generalizable knowledge and to keep subjects as safe as possible under the circumstances. In early-stage regenerative medicine research, the combination of promise, complexity, and uncertainty makes research design and conduct especially challenging.

The consent form and process help to make clear the scientific nature of research goals and the provisional and iterative nature of research progress. The role of informed consent in clinical research is twofold: to promote the autonomy of potential subjects and to encourage critical reflection by researchers and potential subjects (Capron 1974). Yet the best-informed consent is not by itself sufficient to ensure that early-stage research has adequately addressed the ethical issues that accompany research design and conduct; only early, continual, and ongoing critical reflection by researchers can accomplish what is needed to bring science and ethics together in early-stage regenerative medicine research.

Acknowledgments Tristan Keys, MD, Research Resident, Department of Urology, Wake Forest School of Medicine/Wake Forest Institute for Regenerative Medicine, provided invaluable research assistance.

References

Appelbaum, P. S., Roth, L. H., & Lidz, C. (1982). The therapeutic misconception: Informed consent in psychiatric research. *International Journal of Law & Psychiatry, 5*, 319–329.

Atala, A. (2011). *Printing a human kidney, TED talk March 2011*. Retrieved February 23, 2012, from http://www.ted.com/talks/lang/en/anthony_atala_printing_a_human_kidney.html.

Atala, A., Bauer, S. B., Soker, S., Yoo, J. J., & Retik, A. B. (2006). Tissue-engineered autologous bladders for patients needing cystoplasty. *The Lancet, 367*, 1241–1246.

Atala, A., Lanza, R., Thomson, J. A., & Nerem, R. (Eds.). (2010). *Principles of regenerative medicine* (2nd ed.). Amsterdam: Elsevier Science.

Bretzner, F., Gilbert, F., Baylis, F., & Brownstone, R. M. (2011). Target populations for first-in-human embryonic stem cell research in spinal cord injury. *Cell Stem Cell, 8i*, 468–475.

Capron, A. M. (1974). Informed consent in catastrophic disease research and treatment. *University of Pennsylvania Law Review, 123*, 340–438.

Churchill, L. R., Collins, M. L., King, N. M. P., Pemberton, S., & Wailoo, K. A. (1998). Genetic research as therapy: Implications of "gene therapy" for informed consent. *Journal of Law, Medicine & Ethics, 26*, 38–47.

Churchill, L. R., Nelson, D. K., Henderson, G. E., King, N. M. P., Davis, A. M., Leahey, E., et al. (2003). Assessing benefits in clinical research: Why diversity in benefit assessment can be risky. *IRB: Ethics & Human Research, 25*, 1–8.

Cohen, C. B., & Cohen, P. J. (2010a). International stem cell tourism and the need for effective regulation: Part I: Stem cell tourism in Russia and India: Clinical research, innovative treatment, or unproven hype? *Kennedy Institute of Ethics Journal, 20*, 27–49.

Cohen, C. B., & Cohen, P. J. (2010b). International stem cell tourism and the need for effective regulation: Part II: Developing sound oversight measures and effective patient support. *Kennedy Institute of Ethics Journal, 20*, 207–230.

Dresser, R. (2002). The ubiquity and utility of the therapeutic misconception. *Social Philosophy & Policy, 19*, 271–294.

Dresser, R. (2009). First-in-human trial participants: Not a vulnerable population, but vulnerable nonetheless. *Journal of Law, Medicine, & Ethics, 37*, 38–50.

Emanuel, E., Wendler, D., & Grady, C. (2000). What makes clinical research ethical? *Journal of the American Medical Association, 283*, 2701–2711.

Frantz, S. (2012). Embryonic stem cell pioneer Geron exits field, cuts losses. *Nature Biotechnology, 30*, 12–13.

Frey-Vasconcells, J., Whittlesey, K. J., Baum, E., & Feigal, E. G. (2012). Translation of stem cell research: Points to consider in designing preclinical animal studies. *Stem Cells Translational Medicine, 1*, 353–358.

Henderson, G. E., Churchill, L. R., Davis, A. M., Grady, C., Joffe, S., Kass, N., et al. (2007). Clinical trials and medical care: Defining the therapeutic misconception. *PLoS Medicine, 4*, 1735–1738.

Henderson, G., Easter, M. M., Zimmer, C., et al. (2006). Therapeutic misconception in early phase gene transfer trials. *Social Science & Medicine, 62*, 239–253.

Horng, S., & Grady, C. (2003). Misunderstanding in clinical research: Distinguishing therapeutic misconception, therapeutic misestimation, and therapeutic optimism. *IRB: Ethics & Human Research, 25*(1), 11–16.

Ichim, T., Riordan, N. H., & Stroncek, D. F. (2011). The king is dead, long live the king: Entering a new era of stem cell research and clinical development. *Journal of Translational Medicine, 9*, 218–219.

Kimmelman, J. (2004). Valuing risk: The ethical review of clinical trial safety. *Kennedy Institute of Ethics Journal, 14*, 369–393.

Kimmelman, J. (2005). Recent developments in gene transfer: Risk and ethics. *British Medical Journal, 330*, 79–82.

Kimmelman, J. (2009). *Gene transfer and the ethics of first-in-human trials: Lost in translation*. Cambridge: Cambridge University Press.

Kimmelman, J., & London, A. J. (2011). Predicting harms and benefits in translational trials: Ethics, evidence, and uncertainty. *PLoS Medicine*. doi:10.1371/journal.pmed 1001010.

King, N. M. P. (2000). Defining and describing benefit appropriately in clinical trials. *Journal of Law, Medicine, & Ethics, 28*, 332–343.

King, N. M. P. (2003). The line between clinical innovation and human experimentation. *Seton Hall Law Review, 32*, 573–582.

King, N. M. P., & Churchill, L. R. (2008). Assessing and comparing potential benefits and risks of harm. In E. Emanuel, C. Grady, R. A. Crouch, R. K. Lie, F. G. Miller, & D. Wendler (Eds.), *Oxford textbook of clinical research ethics* (pp. 514–526). New York: Oxford University Press.

King, N. M. P., & Cohen-Haguenauer, O. (2008). En route to ethical recommendations for gene transfer clinical trials. *Molecular Therapy, 16*, 432–438.

King, N. M. P., Coughlin, C. N., & Atala, A. (2011). Pluripotent stem cells: In search of the "perfect" source. *Minnesota Journal of Law, Science, & Technology, 12*, 715–730.

King, N. M. P, Coughlin, C., & Furth, M. (2009). Ethical issues in regenerative medicine. *Wake Forest Intellectual Property Law Journal, 9*, 215–237.

King, N. M. P., Henderson, G. E., Churchill, L. R., Davis, A. M., Hull, S. C., Nelson, D. K., et al. (2005). Consent forms and the therapeutic misconception: The example of gene transfer research. *IRB: Ethics & Human Research, 27*(1), 1–8.

Miller, M. (2000). Phase I cancer trials: A collusion of misunderstanding. *Hastings Center Report, 30*(4), 34–43.

National Institutes of Health. (nd). *NIH guidance on informed consent for gene transfer research*. Retrieved February 23, 2012, from http://oba.od.nih.gov/oba/rac/ic/index.html.

Sulmasy, D. P., Astrow, A. B., He, M. K., Sells, D. M., Meropol, N. J., Micco, E., & Weinfurt, K. P. (2010). The culture of faith and hope: Patients' justifications for their high estimates of expected therapeutic benefit when enrolling in early phase oncology trials. *Cancer, 116*, 3702–3711.

Appendix A: Global Embryonic Stem Cell Policies as of 2012

Country	Prohibition of procurement of hESCs from human embryos	Allowing procurement of hESCs from supernumerary human embryos	Prohibition of creation of human embryos for research purposes, including cloning	Allowing creation of human embryos for research purposes, including cloning
EUROPE				
Austria	X		X	
Belgium		X		X
Denmark		X	X	
Germany	X		X	
Greece		X	X	
Finland		X	X	
France		X	X	
Iceland			X	
Ireland	X		X	
Italy	X		X	
Netherlands		X	X	
Norway			X	
Poland	X			
Portugal			X	
Spain		X	X	
Sweden	X			
Switzerland		X		
UK	X			X
ASIA				
China				
India		X		X
Israel		X		X
Japan		X		X
Singapore				X
South Korea		X		X

(continued)

L.F. Hogle (ed.), *Regenerative Medicine Ethics: Governing Research and Knowledge Practices*, DOI 10.1007/978-1-4614-9062-3,

(continued)

Country	Prohibition of procurement of hESCs from human embryos	Allowing procurement of hESCs from supernumerary human embryos	Prohibition of creation of human embryos for research purposes, including cloning	Allowing creation of human embryos for research purposes, including cloning
Taiwan		X	X	
OTHER				
Canada		X	X	
USA				
Mexico			X	

Appendix B: Proposed State Personhood Laws (USA)

State	Description	Outcome
Alabama	Senate Bill 205 (February 2013) defines persons as "all human beings resulting from the union of the male sperm with the female egg either from sexual intercourse or in the case of in vitro fertilization, the fertilized egg or eggs relied on by a physician for implantation in the uterus"	Heard in Senate, referred to Health Committee
Alaska	Ballot initiative, 10NRTL, proposed (2010), defined persons as "all mankind from the beginning of development"	Rejected by Lt Gov and Atty General. State Supreme Court deemed it unconstitutional (2013) when proponent sued to put it on ballot
Arizona	House Bill 2036 (2012): except in a medical emergency, prohibits one from knowingly performing or inducing an abortion on a pregnant woman if the probable gestational age of her unborn child has been determined to be at least 20 weeks. "Gestational age" means the age of the unborn child as calculated from the first day of the last menstrual period of the pregnant woman	Signed into law by governor in April 2012; in May 2013, Ninth Circuit Court of Appeals struck it down as unconstitutional
Arkansas	Act 1032 (February 2013) defines person to include "offspring of human beings from conception until birth"	Passed state Senate and House in March 2013. Awaiting action from Governor
California	Ballot initiative efforts in 2012 to define persons as: "all living human beings from the beginning of their biological development as human organisms …"	Sponsors failed to submit any signatures by 2012 ballot deadline

(continued)

L.F. Hogle (ed.), *Regenerative Medicine Ethics: Governing Research and Knowledge Practices*, DOI 10.1007/978-1-4614-9062-3,

(continued)

State	Description	Outcome
Colorado	Colorado Fetal Personhood Amendment 62 (2010) defined person as "every human being from the beginning of the biological development of that human being"	2010 amendment defeated by 70 % of voters. A similar 2012 ballot initiative was not included on state's ballot
Florida	Personhood amendment ballot initiative (2014) defines person to include "children at every stage of their development, regardless of the method of creation"	Failed to garner the requisite number of signatures to appear on 2012 ballot. Currently seeking signatures for 2014 ballot
Georgia	Amendment SR 420 (2013) defines person as: "all human beings as persons at any stage of development"	Did not pass floor vote, but can be resurrected within next 2 years. Needs legislative approval to be added to 2014 ballot
Iowa	Senate Joint Resolution 10 filed April 2013 defines person as "every person at any stage of development"	Amendment needs to pass both houses and the state Supreme Court to be included on ballot. Similar previous amendment in 2011 failed
Kansas	House Concurrent Resolution 5029 (2012) defined person "from the beginning of the biological development of that human being, including fertilization from beginning of biological development"	Failed to make the 2012 ballot when House Judiciary Committee elected not to have a hearing
Maryland	Personhood amendment HB 1078 introduced to the House (2010) included "every human being from the beginning of the biological development of that human being, including fertilization"	Did not advance further after an unfavorable report by Health and Government Operations
Mississippi	Amendment 26 to define person "from the moment of fertilization, cloning, or equivalent thereof …"	Balloted in 2011; defeated by 59 % of voters
Missouri	Ballot initiative 2010-068 to define person "from beginning of biological development" (2010)	Did not appear on ballot following court challenge
Montana	Constitutional Initiative 108 (2012) defined person as "all human beings at every stage of development, including the stage of fertilization or conception …"	Failed to get the necessary number of signatures to be included on the 2012 ballot
Nevada	Constitutional amendment proposed for 2012 ballot defining person as "every human being"	Failed (2012 and 2010) to obtain the necessary number of signatures for inclusion on ballot
North Dakota	Senate Concurrent Resolution 4009 (2013) expands its definition of person to encompass "every human being at any stage of development …"	Amendment passed in both House and Senate in March 2013; will appear on the state's 2014 ballot

(continued)

(continued)

State	Description	Outcome
Ohio	Personhood constitutional initiative (2013) defines persons as every "human being at every stage of biological development, including fertilization"	Pending--Supporters must collect requisite number of signatures by July 2013 for the amendment to be on the 2013 ballot
Oklahoma	Initiative Petition 395 (2012) defined person as "any human being from the beginning of the biological development of that human being to natural death"	Bill passed in state Senate, subsequently invalidated by state Supreme Court; U.S. Supreme Court declined to hear the case
Virginia	House Bill 1 introduced (2012) defining person to include "unborn children at every stage of development …"	Passed in House, defeated in Senate. Carried over to next session
Wisconsin	Joint Resolution 77 introduced (2011) defining person as "every human being at any stage of development"	Fails to get out of committee but reintroduced in 2013. Both state Legislatures must pass the bill for it to appear on ballot in 2014

Appendix C: Resources

Research Ethics Guidelines

U.S. NIH Research ethics principles: http://clinicalcenter.nih.gov/recruit/ethics.html
U.S. NIH Responsible conduct of research: http://bioethics.od.nih.gov/research-ethics.html
U.S. Office of Research Integrity: http://ori.dhhs.gov/education/products/RCRintro/

Guidelines Specific to Stem Cell Research

U.S. National Academies of Science Stem Cell guidelines: http://www.nap.edu/openbook.php?record_id=4917
California Institute for Regenerative Medicine guidelines: http://cirm.ca.gov/our-funding/stem-cell-regulations-governing-cirm-grants
Canada: http://www.cihr-irsc.gc.ca/e/34460.html
China: http://www.mhlw.go.jp/english/policy/health-medical/medical-care/dl/guidelines.pdf see also http://www.stemcellschina.net/index.php?option=com_content&view=article&id=229%3Achina-regulation-aligns-with-international-standard&catid=13%3Achina-iotech&Itemid=18&lang=en
European Union (by country): http://www.eurostemcell.org/stem-cell-regulations
France: http://www.loc.gov/lawweb/servlet/lloc_news?disp3_l205402748_text
India: http://www.jipmer.edu/stem_cell_guidelines.pdf
ISSCR guidelines for clinical research: http://www.isscr.org/home/publications/ClinTransGuide

L.F. Hogle (ed.), *Regenerative Medicine Ethics: Governing Research and Knowledge Practices*, DOI 10.1007/978-1-4614-9062-3,

General Information Regarding Regenerative Medicine Projections of Market Size and Demand

The Pattison Report (UK): http://www.york.ac.uk/res/sci/events/FinalConfPres/Connolly.pdf

Japan WIPO patent law (in Japanese): http://www.wipo.int/clea/docs_new/pdf/en/jp/jp006en.pdf.

Appendix D: Glossary

AABB	American Association of Blood Banks
ACLU	American Civil Liberties Union
ALLEA	All European Academies
ARM	Alliance for Regenerative Medicine
ASRM	American Society for Reproductive Medicine
AUTM	Association of University Technology Managers
CCD	Charge-coupled device
cGMP	compliant with Good Manufacturing Practice
CIHR	Canadian Institutes of Health Research
CIRM	California Institute for Regenerative Medicine
COC	Certificate of confidentiality
COI	Conflict of interest
CRO	Contract Research Organization
dbGaP	Database of genotypes and phenotypes
dbSNP	Single nucleotide polymorphism database
DHHS	U.S. Department of Health and Human Services
DNA	Deoxyribonucleic acid
DUA	Data use agreement
ECJ	European Court of Justice
EGA	European Genome-Phenome Archive
ES (ESC) cell	Embryonic stem cell, hESC: human embryonic stem cell
ESCRO	Embryonic stem cell research oversight
ESF	European Science Foundation
FACS	Fluorescent antibody cell sorting
FCOI	Financial conflict of interest
FDA	Food and Drug Administration
FTCR	Foundation for Taxpayer and Consumer Rights (now known as Consumer Watchdog)
GATT	General ggreement on tariffs and trade
GEO	Gene expression omnibus

L.F. Hogle (ed.), *Regenerative Medicine Ethics: Governing Research and Knowledge Practices*, DOI 10.1007/978-1-4614-9062-3,

GINA	Genetic Information Nondiscrimination Act
GLP	Good laboratory practices
GSK	GlaxoSmithKline
GWAS	Genome-wide association study
hESCreg	European Human Embryonic Stem Cell Registry
HESC	Human embryonic stem cell
HFEA	Human Fertilization and Embryology Authority
HGP	Human genome project
HHS	(US Department of) Health and Human Services
HIPAA	Health Insurance Portability and Accountability Act of 1996
HIPSCI	Human induced pluripotent stem cell initiative
HPCs	Hematopoietic progenitor cells
hPSC	Human pluripotent stem cell
HSCI	Harvard Stem Cell Institute
IACUCs	Animal Care and Use Committees
IMI	Innovative medicines initiative
IOM	Institute of Medicine
IP	Intellectual property
IPRs	Intellectual property rights
iPS cell (iPSC)	Induced pluripotent stem cell
IRB	Institutional Review Board
ISCBI	International Stem Cell Banking Initiative
ISCR	International Stem Cell Registry
ISCT	International Society for Cell Therapy
ISSCR	International Society for Stem Cell Research
IVF	In-vitro fertilization
JAX	Jackson Laboratory
MOU	Memorandum of understanding
MRC	Medical Research Council
MTA	Material transfer agreement
NAS	National Academy of Sciences
NHLBI	National Heart, Lung and Blood Institute
NIH	National Institutes of Health
NSERC	Natural Sciences and Engineering Research Council
NMR	Nuclear Magnetic Resonance
NPRM	Notice of Proposed Rule-Making
NSCB	National Stem Cell Bank (*Former U.S. NIH-funded stem cell repository*)
OECD	Organization for Economic Co-operation and Development
OLAW	Office of Laboratory Animal Welfare
ORI	Office of Research Integrity
PACT	Production Assistance for Cellular Therapies
PCT	Patent Cooperation Treaty
PHS	Public Health Service
PPPs	Public private partnerships

PubPat	Public Patent Foundation
RCR	Responsible conduct of research
R&D	Research and development
RNA	Ribonucleic acid
SCLD	Stem cell lineage database
SCNT	Somatic cell nuclear transfer
SFI	Significant financial interests
SGC	Structural Genomics Consortium
SNP	Single-nucleotide polymorphism
SRA	Sequence read archive
TRIPS	Agreement on trade related aspects of intellectual property rights
TTO	Technology transfer office
UBMTA	Uniform biological material transfer agreement
UKSCB	United Kingdom Stem Cell Bank
UniProt	Universal Protein Resource
USPTO (PTO)	United States Patent and Trademark Office
WARF	Wisconsin Alumni Research Foundation

Author Biographies

Tania Bubela, PhD, JD, is an Associate Professor in the School of Public Health, University of Alberta. Her research, funded by the Canadian Institutes for Health Research, Genome Canada, and the Stem Cell Network, focuses on intellectual property and health law in biomedical research generally, specifically genomics and stem cell research.

Amritava Das is currently a graduate student in chemical and biological engineering at the University of Wisconsin-Madison. He was a recipient of the Integrating Research Ethics and Scholarship Fellowship in 2012. He did his undergraduate study at the University of Manchester, where he also spent a year at GlaxoSmithKline. He is currently working on the role of canonical Wnt signaling in human pluripotent stem cells.

Linda F. Hogle, PhD, MBA, is Professor of Medical Social Sciences in the School of Medicine and Public Health at the University of Wisconsin-Madison, and a Fellow at the Wisconsin Institutes for Discovery. Her research includes analyses of social, ethical, and regulatory issues in emerging cell-based and biomedical engineering technologies, as well as studies of interdisciplinary scientific work practices and novel organizational platforms. She has served as an advisor to several international research consortia focusing on stem cell and regenerative medicine.

Nancy M.P. King, JD, is Professor, Department of Social Sciences and Health Policy and Wake Forest Institute for Regenerative Medicine, Wake Forest School of Medicine, and Co-Director, Center for Bioethics, Health, and Society and Graduate Program in Bioethics, Wake Forest University. She is a former member of the NIH Recombinant DNA Advisory Committee and a member of the Ethics Committee for the American Society of Gene and Cell Therapy. Much of her scholarship addresses research ethics issues.

Debra J.H. Mathews, PhD, MA, is the Assistant Director for Science Programs for the Johns Hopkins Berman Institute of Bioethics, an Assistant Professor in the Department of Pediatrics, and Affiliate Faculty in the McKusick-Nathans Institute of Genetic Medicine, Johns Hopkins School of Medicine. Her research interests

L.F. Hogle (ed.), *Regenerative Medicine Ethics: Governing Research and Knowledge Practices*, DOI 10.1007/978-1-4614-9062-3,

focus on the intersection of emerging biotechnologies, ethics, and public policy, in particular in the areas of stem cell science, genetics, and neuroscience.

Amrita Mishra, PhD, is a sociologist and postdoctoral researcher in the Department of Public Health Sciences, University of Alberta. Her research examines efforts to enhance the precompetitive space in mouse-model research.

Kevin E. Noonan, PhD, JD, is a scientist, patent lawyer, and founding author of the biotechnology patent law blog, *Patent Docs.* Dr. Noonan received a PhD in molecular biology from Princeton University, and was a National Cancer Institute postdoctoral fellow at the University of Illinois at Chicago where he studied the genetic basis of cancer and drug resistance. He brings this extensive experience to his legal work, representing pharmaceutical and biotechnology companies and university clients on all aspects of intellectual property protection.

Pilar N. Ossorio, JD, is Associate Professor of Law and Bioethics with appointments in the University of Wisconsin Law School and School of Medicine and Public Health; she is also the inaugural Ethics Scholar-in-Residence at the Morgridge Institute for Research. She has published widely in bioethics with a focus on governance of large-scale biomedical research, data sharing, genomics and ethics, and the use of racial categories in research. She has served on numerous policy and advisory committees, including for the National Institutes of Health, the Institute of Medicine, and the US Department of Health and Human Services.

Sean P. Palecek, PhD, is Professor of Chemical and Biological Engineering at the University of Wisconsin-Madison. Dr. Palecek was recently awarded the Cozzarelli prize in biomedical sciences by the National Academy of Sciences for his work on cardiomyocyte differentiation. His lab has developed methods for differentiating human pluripotent stem cells into cardiomyocytes and blood–brain barrier endothelium. His graduate work at MIT under Dr. Douglas Lauffenburger was followed by postgraduate work at The University of Chicago with Dr. Stephen Kron.

David B. Resnik, JD, PhD, is a bioethicist at the National Institutes of Health with expertise in research ethics, including informed consent practices. Dr. Resnik has published more than 200 articles on various topics in philosophy and bioethics and is the author of eight books. He serves on several editorial boards and is an Associate Editor of the journal Accountability in Research. Resnik is also Chair of the National Institutes of Environmental Health Sciences Institutional Review Board.

David Schaffer, PhD, is a Professor of Chemical and Biomolecular Engineering, Bioengineering, and Neuroscience at University of California, Berkeley, where he also serves as the Director of the Berkeley Stem Cell Center. After earning a PhD in Chemical Engineering he did a postdoctoral fellowship in the laboratory of Fred Gage at the Salk Institute for Biological Studies. Dr. Schaffer applies engineering principles to enhance stem cell and gene therapy approaches for neuroregeneration. David Schaffer has received an NSF CAREER Award, Whitaker Foundation Young Investigator Award, and was named a *Technology Review* Top 100 Innovator. He was also awarded the Biomedical Engineering Society Rita Shaffer Young Investigator Award in 2000, the American Chemical Society BIOT Division Young Investigator Award in 2006, and is a Fellow of the American Institute of Medical and Biological Engineering.

Jessica von Reyn, JD, recently completed a law degree at the University of Wisconsin, where she received the Public Interest Law Foundation Scholar award and a Wisconsin Idea Merit Scholarship. She has a B.Sc. from McGill University in anatomy and cell biology and spent 2 years as a laboratory technician at the Vermont Lung Center, where she worked on physiological models of ventilator-induced and acute lung injury.

Peter W. Zandstra, PhD, is I'Anson Professor of Tissue Engineering and Canada Research Chair in Stem Cell Bioengineering at the University of Toronto Institute of Biomaterials and Biomedical Engineering. Dr. Zandstra is a recipient of numerous awards, the most recent being the Till and McCulloch Award (2013). He is a Fellow of the American Institute for Medical and Biological Engineering and the American Association for the Advancement of Science. His research focuses on understanding how complex communication networks between stem cells and their progeny influence self-renewal and differentiation, and how this information can be applied to the design of novel culture technologies capable of controlling cell fate.

Index

L.F. Hogle (ed.), *Regenerative Medicine Ethics: Governing Research and Knowledge Practices*, DOI 10.1007/978-1-4614-9062-3,

J

MIX
Papier aus verantwortungsvollen Quellen
Paper from responsible sources
FSC® C105338

If you have any concerns about our products,
you can contact us on
ProductSafety@springernature.com

In case Publisher is established outside the EU,
the EU authorized representative is:
Springer Nature Customer Service Center GmbH
Europaplatz 3, 69115 Heidelberg, Germany

Printed by Libri Plureos GmbH
in Hamburg, Germany